Composite Electrolyte & Electrode Membranes for Electrochemical Energy Storage & Conversion Devices

Composite Electrolyte & Electrode Membranes for Electrochemical Energy Storage & Conversion Devices

Editor

Giovanni Battista Appetecchi

MDPI • Basel • Beijing • Wuhan • Barcelona • Belgrade • Manchester • Tokyo • Cluj • Tianjin

Editor
Giovanni Battista Appetecchi
ENEA,
SSPT-PROMAS-MATPRO
Technical Unit
Italy

Editorial Office
MDPI
St. Alban-Anlage 66
4052 Basel, Switzerland

This is a reprint of articles from the Special Issue published online in the open access journal *Membranes* (ISSN 2077-0375) (available at: https://www.mdpi.com/journal/membranes/special_issues/electrochemical_energy).

For citation purposes, cite each article independently as indicated on the article page online and as indicated below:

LastName, A.A.; LastName, B.B.; LastName, C.C. Article Title. *Journal Name* **Year**, *Volume Number*, Page Range.

ISBN 978-3-0365-0738-5 (Hbk)
ISBN 978-3-0365-0739-2 (PDF)

© 2021 by the authors. Articles in this book are Open Access and distributed under the Creative Commons Attribution (CC BY) license, which allows users to download, copy and build upon published articles, as long as the author and publisher are properly credited, which ensures maximum dissemination and a wider impact of our publications.

The book as a whole is distributed by MDPI under the terms and conditions of the Creative Commons license CC BY-NC-ND.

Contents

About the Editor vii

Preface to "Composite Electrolyte & Electrode Membranes for Electrochemical Energy
Storage & Conversion Devices" ix

Giovanni Battista Appetecchi
Composite Electrolyte & Electrode Membranes for Electrochemical Energy Storage &
Conversion Devices
Reprinted from: *Membranes* **2020**, *10*, 359, doi:10.3390/membranes10110359 1

João C. Barbosa, José P. Dias, Senentxu Lanceros-Méndez and Carlos M. Costa
Recent Advances in Poly(vinylidene fluoride) and Its Copolymers for Lithium-Ion
Battery Separators
Reprinted from: *Membranes* **2018**, *8*, 45, doi:10.3390/membranes8030045 5

Maria Assunta Navarra, Lucia Lombardo, Pantaleone Bruni, Leonardo Morelli,
Akiko Tsurumaki, Stefania Panero and Fausto Croce
Gel Polymer Electrolytes Based on Silica-Added Poly(ethylene oxide) Electrospun Membranes
for Lithium Batteries
Reprinted from: *Membranes* **2018**, *8*, 126, doi:10.3390/membranes8040126 41

Cataldo Simari, Ernestino Lufrano, Luigi Coppola and Isabella Nicotera
Composite Gel Polymer Electrolytes Based on Organo-Modified Nanoclays: Investigation on
Lithium-Ion Transport and Mechanical Properties
Reprinted from: *Membranes* **2018**, *8*, 69, doi:10.3390/membranes8030069 53

Jahaziel Villarreal, Roberto Orrostieta Chavez, Sujay A. Chopade, Timothy P. Lodge and
Mataz Alcoutlabi
The Use of Succinonitrile as an Electrolyte Additive for Composite-Fiber Membranes in
Lithium-Ion Batteries
Reprinted from: *Membranes* **2020**, *10*, 45, doi:10.3390/membranes10030045 69

Stephen Munoz and Steven Greenbaum
Review of Recent Nuclear Magnetic Resonance Studies of Ion Transport in Polymer Electrolytes
Reprinted from: *Membranes* **2018**, *8*, 120, doi:10.3390/membranes8040120 83

Francisco González, Pilar Tiemblo, Nuria García, Oihane Garcia-Calvo, Elisabetta Fedeli,
Andriy Kvasha and Idoia Urdampilleta
High Performance Polymer/Ionic Liquid Thermoplastic Solid Electrolyte Prepared by Solvent
Free Processing for Solid State Lithium Metal Batteries
Reprinted from: *Membranes* **2018**, *8*, 55, doi:10.3390/membranes8030055 107

Guk-Tae Kim, Stefano Passerini, Maria Carewska and Giovanni Battista Appetecchi
Ionic Liquid-Based Electrolyte Membranes for Medium-High Temperature Lithium
Polymer Batteries
Reprinted from: *Membranes* **2018**, *8*, 41, doi:10.3390/membranes8030041 119

Lucia Mazzapioda, Maria Assunta Navarra, Francesco Trequattrini, Annalisa Paolone,
Khalid Elamin, Anna Martinelli and Oriele Palumbo
Composite Nafion Membranes with $CaTiO_{3-\delta}$ Additive for Possible Applications in
Electrochemical Devices
Reprinted from: *Membranes* **2019**, *9*, 143, doi:10.3390/membranes9110143 133

Maria Montanino, Giuliano Sico, Anna De Girolamo Del Mauro and Margherita Moreno
LFP-Based Gravure Printed Cathodes for Lithium-Ion Printed Batteries
Reprinted from: *Membranes* **2019**, *9*, 71, doi:10.3390/membranes9060071 . . . . **145**

About the Editor

Giovanni Battista Appetecchi is currently a researcher in the Materials and Physicochemical Processes Technical Unit (SSPT-PROMAS-MATPRO) at ENEA (Italian National Agency for New Technologies, Energy and the Sustainable Economic Development). After graduation in Industrial Chemistry at La Sapienza University of Rome on 1993, he was appointed a grant fellowship at La Sapienza University until 2002, and worked as a contract researcher for DAIKIN INDUSTRIES (Japan) and ENEA from 1998 to 2004. In 2004, he was appointed as a researcher at ENEA. In 2013, he spent a one-year sabbatical period in the Department of Chemistry at La Sapienza University in Rome for research activities on materials/components for advanced lithium battery systems. He was the visiting scientist at the Karlsruhe Institute of Technology and the Helmholtz Institute of Ulm (Germany) from 1 September 1 2015 to 2 December 2015; 12 September 2016 to 11 December 2016; and 1 October 2018 to 31 December 2018. For 27 years he has worked on basic and applied research focused on the development of materials and systems for innovative electrochemical energy storage devices, mainly lithium and sodium batteries. His primary research topics are as follows: polymer/gel electrolytes; ionic liquids, composite electrodes. He is the author/co-author of 169 publications in peer reviewed international scientific journals (H-Index = 50, 9,056 citations), 7 book chapters, 163 communications/lectures at scientific meetings, and 3 patents. He is scientifically responsible for ENEA for National and European research projects. In recognition of his work, he was awarded the 1996 "Rotary Club Firenze Est—Herberto Hauda" Award and, in 2012, the Energy and Mobility Award. Dr. Giovani Battista Appetecchi is qualified as a full professor (National Scientific Qualifications accordingly to D.D. 161/2013, January 28, 2013) in the sectors 03/B1 "Principles of Chemistry and Inorganic Systems" and 03/B2 "Chemical Basis of Technology Applications". He is currently Guest Editor of *Membranes* and reviewer of some of the most prestigious international journals in electrochemistry, physical chemistry, science of materials, and power sources.

Preface to "Composite Electrolyte & Electrode Membranes for Electrochemical Energy Storage & Conversion Devices"

Electrolyte membranes have found large employment as separators in electrochemical energy systems, such as flow batteries, fuel cells, electrolyzers, etc., and as such have found increasing applications also in commercial batteries and supercapacitors. One of the most promising approaches for improving the safety and reliability of the commercial rechargeable lithium batteries is the adoption of ionically conducting polymer membranes. These membranes of the capacity to reduce/minimize the liquid leakage within a device. A large effort is currently devoted towards optimizing and improving the chemistry and formulation of the electrolyte separator as it plays a key role in the performance and safety of the electrochemical system. Similarly, the formulation of electrode membranes strongly influences power density, cycling behavior, and reliability. Therefore, these issues are currently under worldwide investigation and represent a key point for the development of the final electrochemical system. The goal of this Special Issue is to present a variegated and a detailed overview on the last findings and frontier approaches regarding these challenging topics. Moreover, it represents an optimal site for welcoming the latest innovations obtained by key laboratories covering the following issues: gel and solvent-free electrolyte membranes for lithium batteries, electrolyte membranes for fuel cells, and electrode membranes for lithium batteries.

Giovanni Battista Appetecchi
Editor

Editorial

Composite Electrolyte & Electrode Membranes for Electrochemical Energy Storage & Conversion Devices

Giovanni Battista Appetecchi

ENEA (Italian National Agency for New Technologies, Energy and Sustainable Economic Development) Materials and Physicochemical Processes Technical Unit (SSPT-PROMAS-MATPRO), Via Anguillarese 301, S. Maria di Galeria, 00123 Rome, Italy; gianni.appetecchi@enea.it; Tel.: +39-06-3048-3924

Received: 9 November 2020; Accepted: 17 November 2020; Published: 21 November 2020

1. Introduction

Currently, electrolyte membranes are largely employed, due to their intriguing peculiarities, as separators in electrochemical energy storage systems such as flow batteries, fuel cells, electrolyzers, etc., and they are also finding applications in commercial batteries and supercapacitors. For instance, one of the most promising approaches for improving safety and reliability of commercial lithium-ion batteries is the adoption of ionically conducting polymer membranes, because of their capability to reduce/minimize the (organic) liquid leakage within the device. Large effort is currently devoted towards optimization and improvement of the chemistry (polymer, salt, solvent, additive, etc.) and formulation of the electrolyte separator as it plays a key role in determining performance and safety of the electrochemical system. Similarly, the formulation of electrodic membranes, which contain passive components such as electronic and/or ionic conductors, polymer binders, etc. not affecting the energy density of the device, strongly influence, however, a device's power density, cycling behavior and reliability. Therefore, although well-known over time, these issues are currently under deep worldwide investigation and represent a key point for the development of an improved electrochemical system.

The goal of the present Special Issue is presenting a variegated and a detailed overview of the latest findings and frontier approaches regarding these challenging topics. Also, it represents an optimal site for welcoming the latest innovations obtained by key laboratories presently involved. The contained articles cover the following highlights: (i) gel electrolyte membranes for lithium battery systems; (ii) solvent-free electrolyte membranes for lithium battery systems; (iii) electrolyte membranes for fuel cells; and (iv) composite electrode membranes for lithium battery systems. A brief descriptive summary of the scientific contributions is reported here.

2. Special Issue Highlights

2.1. Gel Electrolyte Membranes for Lithium Battery Systems

Lithium-ion battery systems are one of the best, if not the best, power sources because of their much higher energy density and cycling performance with respect to other cell technologies. However, they use flammable and volatile organic electrolytes, which represent major safety and reliability concerns, leading to dangerous events (thermal runaway) such as venting, fire and cell bluster. A strategy for increasing their safety level is represented by the confinement of the organic electrolyte within proper polymer hosts for obtaining gel Li^+-conducting membranes able to combine mechanical characteristics of solids (retention of organics) with ion transport properties typical of liquid solutions. Several contributions to this Special Issue focus on this topic.

The manuscript by Barbosa et al. [1] offers a review about recent advancements on different families (homopolymers, co-polymers, composites, and polymer blends) of lithium battery electrolyte separators based on poly(vinylidene fluoride), PVdF. For instance, the separator membrane plays a key role on the battery performance and cycle life, i.e., allowing fast Li$^+$ transport and, at the same time, good compatibility towards both electrodes. In particular, the authors highlight the importance of parameters such as polarity and porosity on the properties of the electrolyte. Finally, a comparison with different types of separator membranes is given.

Several compounds were proposed as additives and/or co-components aiming to improve the characteristics (physicochemical/electrochemical/mechanical properties) of gel electrolyte systems. In the present Special Issue, different approaches (even in terms of technological advances) are presented and discussed by the authors.

Navarra et al. [2] report a physicochemical and electrochemical investigation on a composite gel polymer electrolyte (GPE) based on high molecular weight poly(ethylene oxide), PEO, and polymer host reinforced with nanosized silica. The PEO-SiO$_2$ blend was obtained by the electrospinning technique, allowing production of the composite polymer host as composed by entangled micro-fibers able of housing/retaining the liquid electrolyte. It was then gelled in a mixed solution formed by an organic solvent-ionic liquid-lithium salt solution. Ionic liquids (ILs), i.e., salts molten at room temperature or below, are non-volatile and non-flammable compounds which are proposed as alternative solvents (in place of the organic ones) for enhancing the safety and reliability of lithium batteries.

Simari et al. [3] propose the incorporation of organo-modified montmorillonite clays as nano-additives into a GPE system constituted by poly(acrylonitrile)/poly(ethylene oxide), PAN/PEO, blend (polymer host), lithium salt and organic solvents (ethylene carbonate/propylene carbonate mixture). The organo-clays were prepared by intercalation of CTAB molecules in the interlamellar space of sodium smectite clay through a cation-exchange reaction. A multinuclear NMR spectroscopy investigation allowed measuring the ^{7}Li and ^{19}F self-diffusion coefficients, and the spin-lattice relaxation times. Additionally, a full description of the ions dynamics is reported (including ion transport number and ionicity index).

The effect of ionic liquid addition into GPEs is reported and discussed also in the contribution by Villarreal et al. [4], who have been investigating a mixed organic/ionic liquid electrolytic system. In addition, the authors studied the beneficial effect of incorporation of succinonitrile additive (5 wt.%) on the GPE ion transport properties and on the cycling performance of Li/SnO$_2$-C and Li/LiCoO$_2$ polymer half-cells.

2.2. Solvent-Free Electrolyte Membranes for Lithium Battery Systems

Polymer-salt systems are liquid-free, solid-state solutions (namely solid polymer electrolytes, SPEs) in which the polymer acts as the solvent and the salt (LiX) is the solute. A typical example is represented by the PEO-LiX complexes. The results of development of SPEs are appealing from the safety and technological (i.e., possibility to be manufactured, easily and at low-cost, into thicknesses and shapes not allowed by liquid electrolytes, and better flexibility and mechanical robustness) points of view, thus opening new perspectives to applications in lithium batteries. The absence of any liquid (organic) leakage remarkably enhances the safety and reliability of the electrochemical device and substantially improves the interfacial behavior; these advantages could make possible the employment of lithium metal anodes.

The manuscript by Munoz et al. [5], highlighting how polymer electrolyte development is still a priority, gives a wide review aimed at illustrating various approaches to PEO SPEs, using the NMR spectroscopy technique for investigating their chemical structure and ion transport properties.

PEO electrolyte systems are known to exhibit ion conductivity values suitable for practical applications only at medium-high temperatures (above 70 °C, at which PEO is fully amorphous); this considerably narrows the operative temperature range of all-solid-state lithium polymer batteries. Therefore, a large effort was devoted to prevent PEO host crystallization, thus extending the amorphous phase content at lower temperatures. A very promising approach for improving the ion transport properties of SPEs is represented by the incorporation of ionic liquids into the polymer matrix.

The contributions by González et al. [6] and Kim et al. [7] focus on an IL-containing, thermoplastic PEO electrolyte, prepared through solvent-free processes, to obtain solid, Li$^+$-conducting membrane separators. In addition, González et al. proposed a reinforcement of the SPE system with modified sepiolite (TPGS-S). The authors demonstrate how the incorporation of the ionic liquid largely improves the thermal, ion-transport and interfacial properties (particularly with the lithium metal anode) of the polymer electrolyte. Tests carried out in all-solid-state cells have shown excellent cycling performance and capacity retention, even at high rates, which are never tackled by ionic liquid-free SPEs.

2.3. Electrolyte Membranes for Fuel Cells

Electrochemical energy conversion devices as fuel cells are rather attractive for automotive and stationary applications. The core of this clean technology is the electrolyte separator, typically a Nafion (sulfonate tetrafluoroethylene-based copolymer) proton exchange membrane. For more efficient power generation, fuel cells should operate above 80 °C at which temperature, however, Nafion suffers from water evaporation, leading to remarkable proton conductivity decay. One approach for overcoming this drawback is represented by the incorporation of inorganic additives into Nafion membranes in order to improve the water retention.

In the frame of the present Special Issue, Mazzapioda et al. [8] report a physicochemical investigation into composite proton-conducting membranes for fuel cell applications, based on a Nafion host and incorporating a calcium titanium oxide ($CaTiO_{3-\delta}$) filler. The authors discuss how the addition of the additive, besides enhancing the mechanical properties, improves the water affinity and ionic conductivity. However, high filler contents are seen to play a detrimental effect.

2.4. Composite Electrode Membranes for Lithium Battery Systems

The fourth highlight of this Special Issue is dedicated to composite electrode membranes to be addressed by electrochemical energy devices, in particular lithium batteries. The scientific and technological community involved in this field knows very well the importance of the electrode formulation, i.e., ability of allowing high electronic conductivity and sufficient porosity in combination with good mechanical properties, on the overall battery behavior [1]. At the same time, even the processing route plays a decisive role in determining the electrode performance, resulting in primary importance for certain application. Among these, printed batteries are receiving increasing interest due to the daily growing use of small-size electronic devices, which require very thin-layer batteries for the energy supply. Having this target in mind, industrial gravure printing can represent a proper production technology due to its high speed and quality, and its capability to realize whatever shape electrodes.

The contribution by Montanino et al. [9] proposes the realization of $LiFePO_4$ cathode tapes through gravure printing. This technique, even if it is one of the most appealing for realizing functional layers, has not been investigated in detail until now. In addition, the authors have explored the possibility of employing the printing technique for battery manufacturing.

3. Final Remarks

To summarize, the increasing need of more and more performant (in terms of energy and power density, cycle life and safety) electrochemical devices for storage and conversion of energy is strongly pushing towards the formulation of the new electrolytes. The research on electrolyte separators capable of granting lower cost, safer, more benign and environmentally friendly cells will also expand in view of a large-scale diffusion of hybrid/electric vehicles and delocalized small stationary storage systems. For instance, the electrolyte membrane formulation is expected to play a key role in the cell performance rating capability, compatibility with electrodes, cycling behavior and safety. Also, a deeper understanding of the polymer electrolyte restricted inside the electrode pores is needed for a proper formulation design. In the frame of this scenario, the present Special Issue aims to offer an overview about the different approaches followed to achieve the above-mentioned targets, in particular in the field of electrolyte membranes.

Funding: This research received no external funding.

Acknowledgments: The editor wishes to thank all contributors to this Special Issue for the time and effort devoted for preparing their manuscript.

Conflicts of Interest: The author declares no conflict of interest.

References

1. Barbosa, J.C.; Dias, J.P.; Lanceros-Méndez, S.; Costa, C.M. Recent advances in poly(vinylidene fluoride) and its copolymers for lithium-ion battery separators. *Membranes* **2018**, *8*, 45. [CrossRef] [PubMed]
2. Navarra, M.A.; Lombardo, L.; Bruni, P.; Morelli, L.; Tsurumaki, A.; Panero, S.; Croce, F. Gel polymer electrolytes based on silica-added poly(ethylene oxide) electrospun membranes for lithium batteries. *Membranes* **2018**, *8*, 126. [CrossRef] [PubMed]
3. Simari, C.; Lufrano, E.; Coppola, L.; Nicotera, I. Composite gel polymer electrolytes based on organo-modified nanoclays: Investigation on lithium-ion transport and mechanical properties. *Membranes* **2018**, *8*, 69. [CrossRef] [PubMed]
4. Villarreal, J.; Orrostieta Chavez, R.; Chopade, S.A.; Lodge, T.P.; Alcoutlabi, M. The Use of Succinonitrile as an Electrolyte Additive for Composite-Fiber Membranes in Lithium-Ion Batteries. *Membranes* **2020**, *10*, 45. [CrossRef] [PubMed]
5. Munoz, S.; Greenbaum, S. Review of recent nuclear magnetic resonance studies of ion transport in polymer electrolytes. *Membranes* **2018**, *8*, 120. [CrossRef] [PubMed]
6. González, F.; Tiemblo, P.; García, N.; Garcia-Calvo, O.; Fedeli, E.; Kvasha, A.; Urdampilleta, I. High performance polymer/ionic liquid thermoplastic solid electrolyte prepared by solvent free processing for solid state lithium metal batteries. *Membranes* **2018**, *8*, 55. [CrossRef] [PubMed]
7. Kim, G.-T.; Passerini, S.; Carewska, M.; Appetecchi, G.B. Ionic liquid-based electrolyte membranes for medium-high temperature lithium polymer batteries. *Membranes* **2018**, *8*, 41. [CrossRef] [PubMed]
8. Mazzapioda, L.; Navarra, M.A.; Trequattrini, F.; Paolone, A.; Elamin, K.; Martinelli, A.; Palumbo, O. Composite nafion membranes with CaTiO3−δ additive for possible applications in electrochemical devices. *Membranes* **2019**, *9*, 143. [CrossRef] [PubMed]
9. Montanino, M.; Sico, G.; De Girolamo Del Mauro, A.; Moreno, M. LFP-based gravure printed cathodes for lithium-ion printed batteries. *Membranes* **2019**, *9*, 71. [CrossRef] [PubMed]

Publisher's Note: MDPI stays neutral with regard to jurisdictional claims in published maps and institutional affiliations.

© 2020 by the author. Licensee MDPI, Basel, Switzerland. This article is an open access article distributed under the terms and conditions of the Creative Commons Attribution (CC BY) license (http://creativecommons.org/licenses/by/4.0/).

Review

Recent Advances in Poly(vinylidene fluoride) and Its Copolymers for Lithium-Ion Battery Separators

João C. Barbosa [1,†], José P. Dias [1,†], Senentxu Lanceros-Méndez [2,3,*] and Carlos M. Costa [1,4]

1 Centro de Física, Universidade do Minho, 4710-057 Braga, Portugal; joaocpbarbosa@live.com.pt (J.C.B.); jmpedrodias@gmail.com (J.P.D.); cmscosta@fisica.uminho.pt (C.M.C.)
2 BCMaterials, Basque Center Centre for Materials, Applications and Nanostructures, UPV/EHU Science Park, 48940 Leioa, Spain
3 IKERBASQUE, Basque Foundation for Science, 48013 Bilbao, Spain
4 Centro de Química, Universidade do Minho, 4710-057 Braga, Portugal
* Correspondence: senentxu.lanceros@bcmaterials.net; Tel.: +34-94-612-8811
† Equal contribution.

Received: 29 June 2018; Accepted: 12 July 2018; Published: 19 July 2018

Abstract: The separator membrane is an essential component of lithium-ion batteries, separating the anode and cathode, and controlling the number and mobility of the lithium ions. Among the polymer matrices most commonly investigated for battery separators are poly(vinylidene fluoride) (PVDF) and its copolymers poly(vinylidene fluoride-co-trifluoroethylene) (PVDF-TrFE), poly(vinylidene fluoride-co-hexafluoropropylene) (PVDF-HFP), and poly(vinylidene fluoride-cochlorotrifluoroethylene) (PVDF-CTFE), due to their excellent properties such as high polarity and the possibility of controlling the porosity of the materials through binary and ternary polymer/solvent systems, among others. This review presents the recent advances on battery separators based on PVDF and its copolymers for lithium-ion batteries. It is divided into the following sections: single polymer and co-polymers, surface modification, composites, and polymer blends. Further, a critical comparison between those membranes and other separator membranes is presented, as well as the future trends on this area.

Keywords: PVDF; copolymers; battery separator; lithium-ion batteries

1. Introduction

In the field of mobile applications, the efficient storage of energy is one of the most critical issues, since there is a fundamental need to maximize the amount of energy stored. This issue can be accomplished by increasing the gravimetric and volumetric energy density of the batteries [1].

The electrochemical lithium ion battery is used to provide power to a large variety of mobile appliances, such as smartphones, tablets, and laptops, as well as an increasing number of sensors and actuators, which will have a fundamental role in the shaping of the Internet of Things and Industry 4.0 concepts, the main trend for current technological evolution [2]. Lithium ion batteries can also power electric and hybrid vehicles, and take part in the management of renewable energy production, being essential in a more sustainable energy paradigm. As some renewable resources, such as solar and wind, are intermittent over time, storing energy for their use during periods of lack of resources is a critical issue for lithium ion batteries [3,4].

Lithium ion batteries are very suitable for the aforementioned applications due to their advantages with respect to other battery types, as they are lighter, cheaper, have a higher energy density (250 Wh·kg^{-1}, 650 Wh·L^{-1}), lower charge lost, no memory effect, a prolonged service-life, and a higher number of charge/discharge cycles [5].

Furthermore, the global market of lithium ion batteries is currently growing, and it is expected that in 2022, the market value will reach $46.21 billion, with an annual growth rate of 10.8% [6].

The first commercial lithium ion battery, which was by Sony, entered the market in 1991, with the fundamental contribution of John Goodenough in the development of $LiCoO_2$ as the active material for the cathode [7].

The main components of a battery are the anode, the cathode, and the separator, which are represented in Figure 1, together with the working principle of a lithium ion battery.

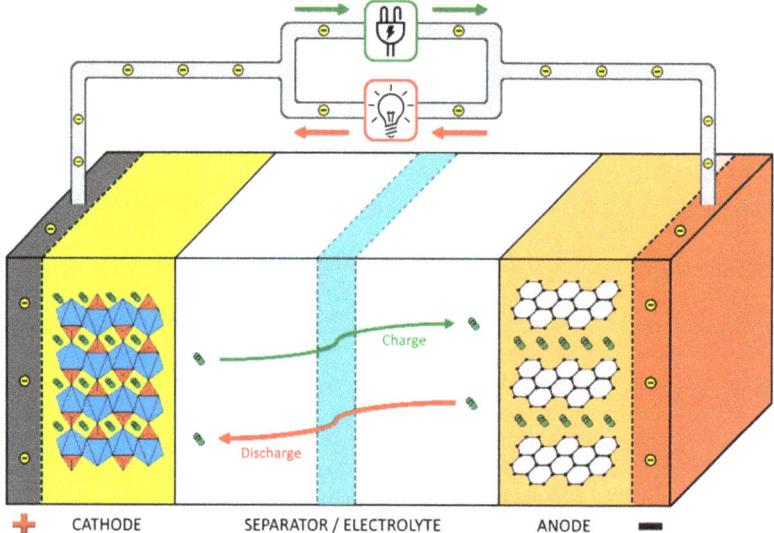

Figure 1. Schematic representation of a lithium ion battery and its working operation.

During the discharge process of the battery, the cathode acts as an oxidizing element, receiving electrons from the external electric circuit and being reduced. The anode is the reducing element, releasing electrons to the external electrical circuit, being oxidized during the electrochemical reaction [8].

2. Battery Separator: Function, Characteristics, and Types

Separators play a key role in the operation of electrochemical devices. The main purpose of the separator membranes is to separate the cathode from the anode, avoiding the occurrence of short circuits, and controlling to the mobility of lithium ions between electrodes. The performance of a separator in a lithium ion battery is determined by some requirements such as porosity, chemical and thermal stability, electrical insulator, wettability, dimensional stability, and resistance to degradation by chemical reagents and electrolytes (Figure 2) [9]. Figure 2 shows the ideal values for the main requirements of a separator membrane.

Figure 2. Ideal values for the main requirements of a separator membrane.

There are different types of separators, but the most widely used consist of a polymer matrix embedded by the electrolyte solution, i.e., a liquid electrolyte where salts are dissolved in solvents, water, or organic molecules. The main types of separators are shown in Table 1 [10].

Table 1. Types and characteristics of different separators adapted from [10].

Separator	Characteristics	Typical Materials
Microporous	Operates at low temperatures (<100 °C); pore size = 50–100 Å	Nonwoven fibers (cotton, nylon, polyester, glass), polymers (PP, PE, PVC, PTFE), rubber, asbestos, wood
Nonwoven	Resistance to degradation by electrolytes; thickness > 25 µm; pore size = 1–100 µm	Polyolefins (PE, PP, PA, PTFE; PVDF; PVC)
Ion exchange membrane	High chemical resistance; impervious to electrolytes; pore size < 20 Å	PE, PP, Teflon-based films
Supported liquid membrane	Solid matrix with a liquid phase; insolubility in electrolyte; high chemical stability	PP, PSU, PTFE, CA
Polymer electrolyte	Simultaneously separator and electrolyte; high chemical and mechanical integrity	Polyethers, PEO, PPO, lithium salts
Solid ion conductor	simultaneously separator and electrolyte	-

The most commonly used materials as matrix for lithium ion battery separators are polymers, or polymer composites. Some of the most commonly used polymers are poly(propylene) (PP), poly(ethylene) (PE), poly(vinylidene fluoride) (PVDF) and its copolymers, poly(ethylene oxide) (PEO), and poly(acrylonitrile) (PAN) [11]. Some separators are developed by blending two different polymers to improve the characteristics of the membrane. In some cases, nanoparticles are added to the matrix as fillers to increase its mechanical stability or ionic conductivity. In composites separators, the most widely used fillers are oxide ceramics (ZrO_2 [12,13], Al_2O_3 [14,15], SiO_2 [16,17]), carbonaceous fillers (graphene [18], carbon black [19], carbon nanofiber [20]), and ionic liquids [21], among others.

The solvents must possess some requirements to ensure proper battery operation. The properties of a good solvent are high dielectric constant, low viscosity, high chemical stability, and in liquid form over a wide temperature range. For this application, solvents of ethylene carbonate (EC), propylene carbonate (PC), dimethyl carbonate (DMC), diethyl carbonate (DEC), and ethyl methyl carbonate (EMC) are the most commonly used [11].

3. Poly(vinylidene fluoride) and Its Copolymers

Considering the different polymer matrices used for battery separators, PVDF and its copolymers (poly(vinylidene fluoride-co-trifluoroethylene), poly(vinylidene fluoride-co-trifluoroethylene) (PVDF-TrFE), poly(vinylidene fluoride-co-hexafluoropropylene), poly(vinylidene fluoride-co-hexafluoropropylene) (PVDF-HFP), and poly(vinylidene fluoride-cochlorotrifluoroethylene) (PVDF-CTFE)) show exceptional properties and characteristics for the development of battery separators, highlighting high polarity, excellent thermal and mechanical properties, wettability by organic solvents, being chemically inert and stable in the cathodic environment, and possessing tailorable porosity through binary and ternary solvent/non-solvent systems [22,23]. The main properties of these polymers are presented in Table 2 [11].

PVDF and its copolymers are partially fluorinated semi-crystalline polymers where the amorphous phase is located between the crystalline lamellae arranged in spherulites. It can crystallize in different crystalline phase, depending on the temperature and processing conditions [24,25]. In relation to the crystalline phases of PVDF and its copolymers, the most important phases are the β-phase, since it presents ferroelectric, piezoelectric, and pyroelectric properties, and the α-phase, which is the most stable thermodynamically, when material is obtained directly from the melt [24]. As illustrated in Table 2, PVDF and its polymers are characterized by excellent mechanical properties, good thermal stability up to 100 °C, and a high dielectric constant, which is essential for assisting the ionization of lithium salts.

Table 2. Main properties of PVDF and its copolymers [26–28].

Polymer	Melting Temp./°C	Degree of Crystallinity/%	Young Modulus/MPa	Dielectric Constant
PVDF	~170	40–60	1500–3000	6–12
PVDF-TrFE	~120	20–30	1600–2200	18
PVDF-HFP	130–140	15–35	500–1000	11
PVDF-CTFE	~165	15–25	155–200	13

PVDF copolymers have drawn increasing attention for battery separators, as the addition of other monomers to the VDF blocks increases the fluorine content and decreases the degree of crystallinity (Table 2), which is particularly relevant once the uptake of the electrode solution occurs in the amorphous region through a swelling process for accommodating the electrolyte and, as a result, increases the ionic conductivity [29]. The recent literature on PVDF and its battery separator copolymers is structured into four sections dedicated to single polymers, surface modification, composites, and polymer blends, respectively.

The main achievement for PVDF and co-polymers as battery separators was thoroughly reviewed in [11]. Since then, important contributions have been achieved, which are the subject of the present review.

3.1. Single Polymer and Co-Polymers

As already mentioned, one of the main characteristics of PVDF and its co-polymers is their high dielectric permittivity, providing a large affinity with polar electrolytes when compared to other polymers [11]. The main characteristics of the developed PVDF and the copolymer membranes are shown in Table 3.

Table 3. Separator membranes based on PVDF and co-polymers, indicating also the main properties, and the main goal/achievement of the investigation.

Materials	Electrolyte Solution	Porosity and Uptake (%)	Conductivity (S·cm^{-1}) and Capacity (mAh·g^{-1})	Main Goal/Achievement	Reference
PVDF	1 M (C$_2$H$_5$)$_3$CH$_3$NBF$_4$ + AN	-/-	-/-	Study of multistep electrospinning technique on the fabrication of PVDF composite membranes; High specific power.	[30]
PVDF	1 M LiPF$_6$ in EC:DEC (1:1, w/w)	-/816	6.83 × 10^{-4}/101.1 (0.5C)	Performance comparison with a PVDF-PDA separator; Enhanced cycling performance.	[31]
PVDF	1 M LiPF$_6$ in EC:DEC (1:1, v/v)	7/-	-/-	Analysis of the migration mechanism of cation and anions through the separator; The separator allows the control of structural stability and ion mobility.	[32]
PVDF	1 M LiPF$_6$ in EC/DMC/EMCC (1:1:1, $w/w/w$)	-/-	-/95 (0.2C)	Production of a PVDF membrane; Good capacity retention.	[33]
PVDF	1 M TEABF$_4$ in AN/PC and 1 M LiPF$_6$ in EC/DEC	80/-	1.8 × 10^{-2} (25 °C)/-	Manufacturing of a PVDF separator; Favorable mechanical properties.	[34]
PVDF	1 M LiBF$_4$ in EC/DMC (50:50 wt. %)	-/-	4.17 × 10^{-3} (20 °C)/-	Comparison of PVDF membrane performance with Nafigate separators.	[35]
PVDF	1 M LiPF$_6$ in EC/DMC/DEC (1:1:1)	78.9/427	1.72 × 10^{-3}/164.3 (C/5)	Synthesis of dual asymmetric structure separators; Improved electrolyte uptake and ionic conductivity.	[36]
PVDF	1 M LiPF$_6$ in EC/DMC/DEC (1:1:1)	-/-	-/447.36 (0.3C)	Production of a solid state SCPC with a PVDF separator; High storage capacity and stability.	[37]
PVDF	-	-/-	-/-	Assembly of a PVDF separator for air-cathode as application in microbial fuel cells; Improved electricity generation.	[38]
PVDF	PVA/H$_2$SO$_4$	-/-	-/-	Production of a PVDF separator for piezoelectric supercapacitors; High mechanical strength and elevated capacitance.	[39]
PVDF	1 M NaClO$_4$ in EC/DEC (1:1)	81/34	7.38 × 10^{-4} (29 °C)/153	Production of an electroactive electrospun PVDF separator for sodium ion batteries.	[40]

Table 3. Cont.

Materials	Electrolyte Solution	Porosity and Uptake (%)	Conductivity (S·cm^{-1}) and Capacity (mAh·g^{-1})	Main Goal/Achievement	Reference
PVDF	1 M LiPF$_6$ in EC/DEC (1:1)	70/66	1.5×10^{-3}/102 (2C)	Study of the effect of different PVDF copolymers as lithium ion battery separators. Demonstration of the relevance of β-phase content.	[41]
PVDF-TrFE	1 M LiPF$_6$ in EC/DEC (1:1)	72/84	1.1×10^{-3}/118 (2C)		
PVDF-HFP	1 M LiPF$_6$ in EC/DEC (1:1)	56/79	1.3×10^{-1}/107 (2C)		
PVDF-CTFE	1 M LiPF$_6$ in EC/DEC (1:1)	59/80	1.5×10^{-3}/85 (2C)		
PVDF	[C$_2$mim][NTf$_2$]	20/98	2.3×10^{-4} (25 °C)/74.6 (C/5)	Preparation of PVDF separators using a green solvent and ionic liquid as the electrolyte.	[9]
PVDF-HFP	LiTFSI	48/248	5.2×10^{-5} (20 °C)/-	Application of disiloxane-based electrolytes on PVDF-HFP for the production of gel electrolyte separators; Good thermal and mechanical stability.	[42]
PVDF-HFP	LiNfO/BMImNfO	-/-	2.61×10^{-2}/(100 °C) 138.1 (C/4)	Production of ionic liquid gel polymer electrolytes; High ionic conductivity.	[43]
PVDF-HFP	1 M LiPF$_6$ in EC/DMC (1:2)	70/247	3.2×10^{-3} (25 °C)/-	Evaluation of the performance of PVDF-HFP, as a single polymer membrane. Understanding of the method of avoiding the formation of beads in the nanofibers of PVDF-HFP; Good electrolyte uptake.	[44]
PVDF-HFP	1 M LiPF$_6$ in EC:DMC (1:1)	78/86.2	1.03×10^{-3}/145 (0.2C)	Development of a PVDF-HFP gel polymer electrolyte membrane with honeycomb type porous structure; Excellent electrochemical performance.	[45]
PVDF-HFP	1 M LiPF$_6$ in EC/DEC/EMC (1:1:1)	-/-	-/-	Production of separators with controlled pore structure; Improved rates and cycling performances.	[46]
PVDF-CTFE	1 M LiPF$_6$ in EC:DMC:EMC (1:1:1, v:v)	74/-	7.51×10^{-4} (25 °C)/147 (0.2C)	Preparation of a nanofiber-coated composite separator by electrospinning; High discharge capacity and good cycling stability.	[47]

Table 3 shows that the electrospinning technique is widely used to produce functional membranes. Thus, electrospun separators have been developed for PVDF-PDA [31], PVDF-HFP [44], and PVDF-CTFE [47].

For the PVDF-CTFE membrane, the cell assembly considered for the battery performance tests is represented in Figure 3.

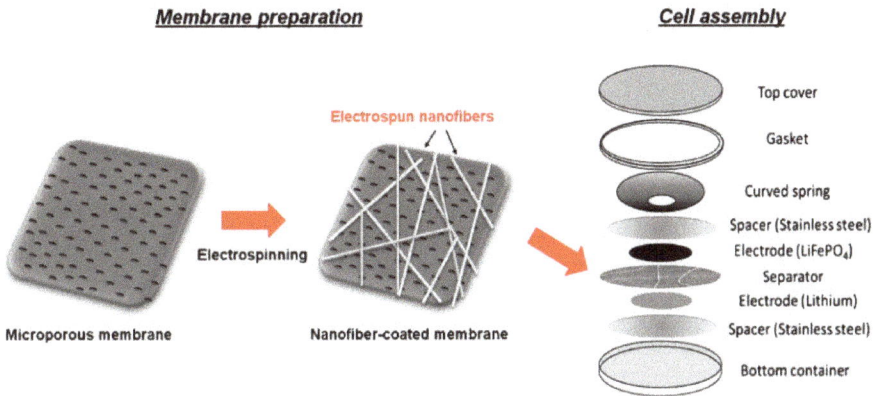

Figure 3. Manufacturing of a testing cell based on PVDF-CTFE separators [47], with copyright permission from Springer Nature.

For PVDF-HFP electrospun membranes, it has been demonstrated that a single layer membrane shows good porosity and uptake value, but that the mechanical stability is negatively affected, with the viscosity of the solution playing an important role [44]. Also, a novel gel electrolyte was developed based on PVDF-HFP by the addition of disiloxane into the electrolyte solution [42], leading to a thermally stable separator that is not flammable, thus contributing to safer lithium ion batteries [45]. It this sense, ionic liquids have also been used in electrolyte solutions, improving both safety and the ionic conductivity of the membranes [43].

A multistep electrospinning technique for the production of PVDF membranes for electrical double-layer capacitors has been proposed, allowing for the manufacture of thinner and more densely packed separators [30].

Further, membranes have been developed based on PVDF for air-cathode in microbial fuel cells [38] and piezo-supercapacitors [39]. Dual asymmetric PVDF separators were produced by a thermally-induced phase separation method, in which the large and interconnected pores in the bulk structure ensures an improved electrolyte uptake and ionic conductivity, while the small pores in the surfaces prevent the loss of electrolyte and the growth of lithium dendrites. It is indicated that those separators ensure safer batteries with high discharge capacity and longer cycle life [36].

A further step towards the development of more environmentally friendly PVDF separator membranes was proposed by using DMPU as a solvent for PVDF, and IL [C2mim][NTf2] as an electrolyte. The use of the IL increased the ionic conductivity and discharge capacity of the membrane when compared with separators using conventional electrolytes [9].

Porous PVDF-HFP membranes were prepared with non-solvents using the phase inversion technique. When selecting different types of non-solvents such as water, methanol, ethanol, and propanol, and their contents in acetone, it was possible to control the size of the pores (Figure 4) [46].

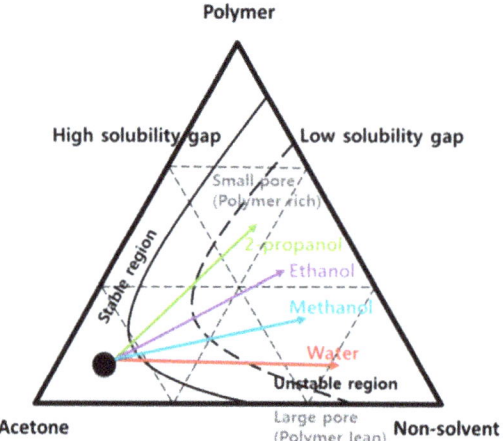

Figure 4. Phase diagram of the ternary mixture—PVDF–HFP, acetone, and non-solvent—in order to control PVDF-HFP membrane morphology [46], with copyright permission from the Royal Society of Chemistry.

Finally, a correlation between the β-phase content of the separators, and the rate capability and cyclability of the batteries was demonstrated for different PVDF co-polymers, showing that the PVDF-TrFE membrane has the best battery performance for the highest β-phase content (100%) [41].

Thus, it is observed that for single (co)polymer membranes, the main focus is to tailor the morphology to obtain good uptake without mechanical deterioration, and to improve the interaction between the electrolyte solution and the separator membrane.

3.2. Surface Modification of the Separator Membranes

Typically, surface modification of the membranes is carried out to improve specific properties such as wettability, and thermal and mechanical stability. PVDF membranes have been prepared after different surface modifications, but also have been used to modify the properties of other polymer membranes, as presented in Table 4.

Table 4. Surface modifications on PVDF and co-polymers, indicating also the main properties, goal and achievement.

Materials	Electrolyte Solution	Porosity and Uptake (%)	Conductivity (S·cm^{-1}) and Capacity (mAh·g^{-1})	Main Goal/Achievement	Ref
PVDF (plasma-treated)	1 M LiPF$_6$ in EC/DMC (1:1)	-/1200	-/-	Study of the effect of plasma treatment in PVDF separators; Improved electrolyte uptake and mechanical properties.	[48]
PE/PVDF	1 M LiPF$_6$ in EC:EMC:DEC (1:1:1, w:w:w)	-/-	0.89×10^{-3} (25 °C)/-	Investigation into the pore formation process in a coating layer for separators; Enhanced ionic conductivity.	[49]
PE/PVDF	1.10 M LiPF$_6$ in EC/PC/EP (3:1:6, v:v:v)	-/-	-/1436 (0.2C)	Study of the electrochemical performance of PE/PVDF separators; Enhanced cycling performance.	[50]
PVDF/PP	1 M LiPF$_6$ in EC/DMC (1:1)	58/140	5.9×10^{-4}/145 (0.5C)	Coating of PVDF particles in the surface of a PP membrane; Increased electrolyte uptake.	[51]
PET/PVDF	1 M LiPF$_6$ in EC/DEC/DMC (1:1:1, w/w/w)	-/-	8.36×10^{-3}/-	Investigation of the performance of a hot-pressed PET/PVDF separator; Excellent mechanical behavior.	[52]
PVDF/HEC	1 M LiPF$_6$ in EC/DMC/EMC (1:1:1)	-/135.4	8.8×10^{-4} (25 °C)/140	Preparation of a PVDF/HEC/PVDF membrane with a sandwich structure; High electrolyte uptake and ionic conductivity.	[53]
PVDF/PMMA	1 M LiTFSI in DME/DOL (1:1)	-/294	1.95×10^{-3} (25 °C)/1711.8	Preparation of a sandwiched GPE based on PVDF and PMMA for lithium-sulfur batteries; High discharge capacity and cycle stability.	[54]
PDA/PVDF	1 M LiPF$_6$ in EC:DEC (1:1, wt:wt)	-/1160	9.62×10^{-4}/104.5 (0.5C)	Prove that the PDA coating can be promising for manufacturing electrospun nanofiber separators; Better cycling performance and elevated power capability.	[31]
PE/(PVDF/Al$_2$O$_3$)	1 M LiPF$_6$ in EC/DEC (1:1)	60.3/125, 314	1.14–1.23×10^{-3}/-	Development of a multilayer coating for separators; Improvement of thermal stability and electrolyte wetting.	[55]
PI/PVDF/PI	1 M LiPF$_6$ in EC/DEC/DMC (1:1:1)	83/476	3.46×10^{-3}/114.8 (0.5C)	Production of an electrospun sandwich-type separator; Superior porosity, electrolyte uptake, and ionic conductivity.	[56]
PVDF-HFP	1 M NaClO$_4$ in EC/PC (1:1)	-/-	3.8×10^{-3}/291.1 (0.2C)	Development of a PVDF-HFP-coated GF separator for sodium ion batteries; Good cycling performance.	[57]

Table 4. Cont.

Materials	Electrolyte Solution	Porosity and Uptake (%)	Conductivity (S·cm^{-1}) and Capacity (mAh·g^{-1})	Main Goal/Achievement	Ref
PVDF-HFP	1 M LiPF$_6$ in DMC/EMC/EC (1:1:1)	53.5/106.9	8.34 × 10^{-4}/131.33 (5C)	Study of the effect of the drying temperature on the performance of the separator.	[58]
PP/(PVDF-HFP/SiO$_2$)	1 M LiPF$_6$ in DEC/EC (1:1, v/v)	-/-	7.2 × 10^{-4}/-	Analysis on the effect of a PVDF-HFP/SiO$_2$ coating layer for PP separators; Better electrolyte uptake and ionic conductivity.	[59]
PMMA/PVDF-HFP	1 M LiPF$_6$ in EC:DMC (1:1)	-/342	1.31 × 10^{-3}/143 (0.2C)	Investigation and analysis of a produced PMMA/PVDF-HFP electrolyte membrane; Exceptional thermal and electrochemical stability.	[60]
PVDF-HFP/PDA	LiPF$_6$ in EC/DEC/DMC (1:1:1)	72.8/254	1.40 × 10^{-3} (20 °C)/-	Production of a PVDF-HFP/PDA separator by a dip-coating method.	[12]
PVDF-HFP/PET	1 M LiClO$_4$ in DMSO	-/282	6.39 × 10^{-3} (25 °C)/158 (0.1C)	Combination of PVDF-HFP with a SiO$_2$ nanoparticle-modified PET matrix; Improved thermal stability, electrolyte uptake, and ionic conductivity.	[61]
PP/(AlO$_2$/PVDF-HFP)	1 M LiPF$_6$ in EC/DEC (1:1, v/v)	-/-	7.95 × 10^{-4}/98.6 (0.2C)	Inspection of the performance of a separator for PP membrane coating; Improved thermal stability.	[62]
γ-Al$_2$O$_3$/PVDF-HFP/TTT	1 M LiClO$_4$ in EC/DEC (1:1)	-/157	1.3 × 10^{-3}/~100 (0.5C)	Dip coating of a PE separator with γ-Al$_2$O$_3$/PVDF-HFP/TTT; Increased electrolyte uptake and ionic conductivity.	[13]
PP/PE/PP/PVDF-co-CTFE	1 M LiPF$_6$ in EC/DMC/DEC (1:1:1, $v:v:v$)	-/-	-/-	Fabrication of PVDF-co-CTFE nanofiber coatings for improving the performance of polyolefin separators; High electrolyte uptake and good wettability.	[63]

The most commonly used surface modification is the use of PVDF and its copolymers for the coating of other polymers such as polyethylene porous separators. Thus, the coating of PE with a Al_2O_3 ceramic layer and a PVDF electrospun nanofiber layer leads to enhanced electrolyte uptake, improved capacity discharge, and cycle life [55]. Similarly, a PDA coating on PVDF improves hydrophilicity, enhancing electrolyte uptake and ionic conductivity of the separator [31].

A typical surface modification technique, such as plasma treatment, allows significant improvement of the electrolyte uptake of PVDF electrospun membranes [48].

A hot-pressing technique was proposed to develop PET/PVDF separators, with improved mechanical behavior properties [52].

The preparation of a PVDF/PMMA/PVDF separator showed great potential for its use in lithium-sulfur batteries, showing high initial discharge capacity and cycle stability, also reducing cell polarization and suppressing the shuttle effect, which is described as the transport of soluble polysulfides between both electrodes and the associated charge [54].

A composite membrane with a PVDF/HEC/PVDF sandwich structure was developed, leading to higher electrolyte uptake, ionic conductivity, and cycling performance. It is also greener and safer because of the fire-retardant behavior of its components [58].

For PVDF-HFP membranes, several coatings have been applied, such as ZrO_2 nanoparticles [64], PP polymer [59], PMMA polymer [60], PDA layer [12], and SiO_2-modified PET [61], leading mainly to improved electrolyte uptake.

Surface modifications are also achieved by modifying the drying temperature of PVDF-HFP/PET separators prepared by dip-coating, with a drying temperature of 80 °C improving cycle and rate performances with respect to batteries with a conventional PP separator [58].

The dip-coating of a PE separator with γ-Al_2O_3/PVDF-HFP/TTT, proved to increase electrolyte uptake and ionic conductivity when compared with conventional membranes, as shown in Figure 5 where its microstructure and cycling performance are presented. The discharge performance was also enhanced as well as the thermal resistance [13].

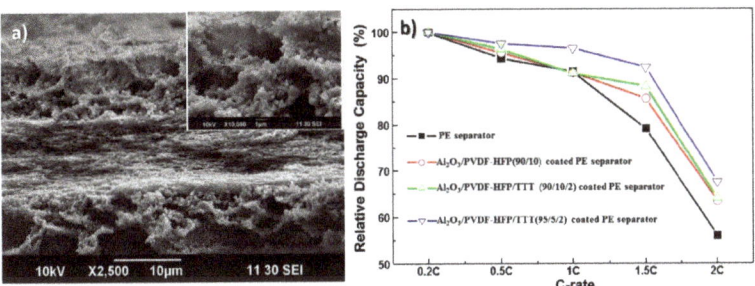

Figure 5. (a) Cross-section scanning electron microscopy (SEM) images of the γ-Al_2O_3/PVDF-HFP/TTT(95/5/2)- coated PE separator and (b) relative discharge capacities as a function of the C-rate [13], with copyright permission from Elsevier.

Basically, surface modifications are essential for improve the electrolyte wettability of the separators, and are realized in several polymer membranes of single and multiple layers with many polymers (PP, PET, PMMA, etc.) and filler nanoparticles.

3.3. Composite Membranes

Polymer composites are used to improve battery performance by incorporating suitable fillers, such as oxides ceramic, zeolites, and carbon nanotubes, among others, with the objective of increasing ionic conductivity, mechanical strength, and thermal stability. The main properties of composite separator membranes based on PVDF and its copolymers are presented in Table 5.

Table 5. Polymer composites based on PVDF and co-polymers with main properties, goal, and achievement.

Materials	Fillers	Electrolyte Solution	Porosity and Uptake (%)	Conductivity (S·cm^{-1}) and Capacity (mAh·g^{-1})	Main Goal/Achievement	Ref.
PVDF	Al_2O_3	1 M $LiPF_6$ in EC/DEC/DMC (1:1:1)	55.8/153.5	2.23×10^{-3} (25 °C)/114.2	Production of a composite PVDF/Al_2O_3; High thermal stability and ionic conductivity, low discharge capacity decay.	[15]
PVDF	Al_2O_3	EC/DMC (1:1)	-/230	1.24×10^{-3}/151.97 (C)	Core-shell composite nonwoven separator of PVDF-HFP@Al_2O_3; high heat resistance up to 200 °C without any shrinkage.	[65]
PVDF	Al_2O_3	1 M $LiPF_6$ in EC/DEC (1:1, v:v)	67/230	1.49×10^{-3}/146.3 (0.2C)	Separator-cathode assembly with PVDF/Al_2O_3; Good electrochemical performance.	[66]
PVDF	AlO(OH) nanoparticles	1 M $LiPF_6$ in EC/DEC (3:7)	-/65	-/-	Ceramic separator based on boehmite nanoparticles; Improved safety and wettability.	[67]
PVDF	BC	1 M LiTFSI in EC/DEC (1:1)	-/-	4.2×10^{-3} (30 °C)/-	Preparation of GPEs based on cross-linkers; High ionic conductivity and thermal stability.	[68]
PVDF	Carbon	1 M LiTFSI and 0.1 M $LiNO_3$ in DOL/DME (1:1)	-/-	-/827 (0.5C)	PVDF-C separator by phase inversion technique; Superior rate performance and stability.	[69]
PVDF	CNF	1 M LiTFSI in DOL/DME (1:1)	-/119	-/1739.2 (C)	Production of CNF/PVDF separators for Li-S batteries Great battery discharge capacity and cycling stability.	[20]
PVDF	Cellulose acetate/Al(OH)$_3$	1 M $LiPF_6$ in EC/DMC/EMC (1:1:1)	68.6/403.9	2.85×10^{-3}/151.97 (C)	Environmental friendly materials in a separator; High electrolyte uptake, ionic conductivity and cycling performance.	[70]
PVDF	DNA-CTMA	$LiAsF_6$ in EC/EMC/DMC	-/-	-/-	PVDF/DNA-CTMA membrane as a solid polymer/gel electrolyte separator; Improved thermal and mechanical properties.	[71]
PVDF	LiPVAOB	1 M $LiPF_6$ in EC/DMC/EMCC (1:1:1, w:w:w)	-/88.5	2.6×10^{-4}/120 (0.2C)	Composite gel polymer electrolyte PVDF/LiPVAOB membrane; Good ionic conductivity.	[33]
PVDF	Nanoclays/PVP	1 M $LiPF_6$ in EC/DMC (1:1)	87.4/553.3	-/-	Study of the influence of solvents in the separator High porosity and uptake.	[72]

Table 5. Cont.

Materials	Fillers	Electrolyte Solution	Porosity and Uptake (%)	Conductivity (S·cm^{-1}) and Capacity (mAh·g^{-1})	Main Goal/Achievement	Ref.
PVDF	NCC	1 M LiPF$_6$ in EC/DMC (1:1)	-/-	3.73×10^{-3} (25 °C)/-	Preparation of NCC-PVDF separators by phase inversion; Improved wettability and mechanical properties.	[73]
PVDF	MA groups	1 M LiPF$_6$ in EC/DMC/EMC (1:1:1)	67.4/-	$1.48 \times 10^{-3}/136$ (0.2C)	Study of the addition of MA groups to the PVDF structure; High ionic conductivity.	[74]
PVDF	MMT	1 M LiPF$_6$ in EC/EMC/DEC (1:1:1)	84.08/333	4.20×10^{-3} (25 °C)/144	Effect of different contents of MMT filler in PVDF separators; High ionic conductivity and porosity.	[75]
PVDF	MOF-808	-	-/-	1.56×10^{-4} (65 °C)/-	Production of a MOF/polymer membrane; Good mechanical properties and durability.	[76]
PVDF	Octaphenyl-POSS	1 M LiPF$_6$ in EC/DMC/EMC (1:1:1)	66.1/912	$4.2 \times 10^{-3}/145.8$ (0.5C)	Electrospun membrane with octaphenyl-POSS particles; Increased uptake and porosity, high ionic conductivity.	[77]
PVDF	Polyether (PEGDA+PEGMEA)	1 M LiPF$_6$ in EC/DMC/EMC (1:1:1)	-/230	~1.4×10^{-3} (25 °C)/93 (0.5C)	Preparation of GPEs with PVDF and polyethers.	[78]
PVDF	PMIA	1 M LiPF$_6$ in EC/DMC/EMC (1:1:1, w:w:w)	-/-	$8.1 \times 10^{-4}/135.29$ (0.2C)	Composite sandwich type separator, by electrospinning; High capacity retention and good rate performance.	[79]
PVDF	P-PAEK	1 M LiPF$_6$ in EC/DMC (1:1)	71.7/123.7	/141.6 (C/2)	Development of a P-PAEK/PVDF separator High wettability and electrolyte uptake.	[80]
PVDF	PFSA	1 M LiPF$_6$ in EC/DMC/EMC (1:1:1)	-/-	$1.53 \times 10^{-3}/137.9$ (C)	PVDF/PFSA blend membrane; High stability and discharge capacity.	[81]
PVDF	rGO	1 M LiTFSI + 0.1 M LiNO$_3$ in DME/DOL (1:1)	71/380	/646	Double-layer PVDF/rGO membrane by electrospinning; High safety and cycling stability.	[82]
PVDF	SiO$_2$	1 M LiPF$_6$ in EC/DMC/EMC (1:1:1)	54.1/279.5	-/175.7	Synthesis of a composite separator with SiO$_2$; High wettability, uptake and thermal/mechanical stability.	[17]
PVDF	SiO$_2$	1 M LiPF$_6$ in EC/EMC (1:1 in volume)	70/370	$2.6 \times 10^{-3}/132$ (C)	Addition of SiO$_2$ nanoparticles on PVDF membranes; Improvement of wettability and ionic conductivity.	[83]
PVDF	SiO$_2$	1 M LiPF$_6$ in EC/DEC (1:1, v:v)	85/646	$7.47 \times 10^{-3}/159$ (0.2C)	Electrospun PVDF/SiO$_2$ composite separator; Excellent thermal stability and high ionic conductivity.	[84]
PVDF	SnO$_2$	1 M LiPF$_6$ in EC/DMC (1:1 w:w)	-/-	-/-	Use of SnO$_2$ nanoparticles in a PVDF electrospun separator; Good cycling performance.	[85]

Table 5. Cont.

Materials	Fillers	Electrolyte Solution	Porosity and Uptake (%)	Conductivity (S·cm^{-1}) and Capacity (mAh·g^{-1})	Main Goal/Achievement	Ref.
PVDF	ZnO	1 M LiPF$_6$ in EC/EMC (1:2)	-/-	-/-	Piezo-separator for integration on a self-charging power cell; Enhanced electrochemical performance.	[86]
PVDF	ZnO	1 M LiPF$_6$ in EC/DEC (1:1)	-/-	-/-	Piezo-separator for self-charging power cells; Stable and efficient performance.	[87]
PVDF	ZrO$_2$/PEO	1 M LiTFSI in DOL/DME (1:1)	-/147.3	3.2×10^{-4} (25 °C)/1429 (0.2C)	GPE for lithium-sulfur batteries; High discharge capacity and rate performance.	[88]
PVDF-HFP	Al$_2$O$_3$	0.5 M NaTf/EMITf	-/-	6.3–6.8×10^{-3} (25 °C)/-	Introduction of Al$_2$O$_3$ in a gel polymer electrolyte; Improved mechanical properties.	[14]
PVDF-HFP	Al$_2$O$_3$	1 M LiPF$_6$ in EC/DEC +2% VC	-/372	1.3×10^{-3}/155 (0.5C)	Colloidal Al$_2$O$_3$ composite separator; enhancement of the mechanical strength of the PVDF-HFP separator.	[89]
PVDF-HFP	Al$_2$O$_3$	1 M LiPF$_6$ in EC/DMC/EMC (v:v:v = 1:1:1)	-/420	4.7×10^{-4}/109 (4C)	Production of a low cost membrane, with a simple and easy scalable manufacturing process; High electrolyte uptake and good electrochemical stability and performance.	[90]
PVDF-HFP	Al(OH)$_3$	1.15 M LiPF$_6$ in EC/EMC (3:7, v:v)	84/127	10^{-3}/81 (C/2)	Upgrading the battery safety operation by the addition of metal hydroxides in composite separators; Suitable electrolyte uptake.	[91]
PVDF-HFP	Al$_2$O$_3$/CMC	1 M LiPF$_6$ in EC/DEC/PC/EMC (2:3:1:3)	42.7/-	9.3×10^{-4} (25 °C)/-	Composite separator with Al$_2$O$_3$/CMC; Safer and more stable separators.	[92]
PVDF-HFP	BN	1 M LiPF$_6$ in EC/DEC (1:1)	-/-	-/150 (0.2C)	3D separator; improved cycling stability with lower voltage polarization	[93]
PVDF-HFP	CA	1 M LiPF$_6$ in EC/DMC	85/310	1.89×10^{-3}/136 (8C)	Porous and honeycomb-structured membrane; higher lithium-ion transference number and improved rate performance	[94]
PVDF-HFP	Clay	1 M LiPF$_6$ in EC/DEC/EMC (1:1:1, v:v:v)	-/-	1.49×10^{-3}/-	New technique to incorporate clay sheets in a PVDF-HFP matrix, as separator; Thermal stability and higher ionic conductivity.	[95]
PVDF-HFP	EMImNfO-LiNfO	-	-/-	3.92×10^{-4}/(20 °C) 57 (C)	Introduction of anion-based IL and lithium salt in a GPE; High thermal stability, good electrochemical properties.	[96]
PVDF-HFP	GO	1 M LiPF$_6$ in EC/DEC/EMC (1:1:1)	-/71	1.115×10^{-3} (25 °C)/-	Addition of GO in separators to increase thermal properties; improved electrochemical and mechanical properties.	[97]

Table 5. Cont.

Materials	Fillers	Electrolyte Solution	Porosity and Uptake (%)	Conductivity (S·cm^{-1}) and Capacity (mAh·g^{-1})	Main Goal/Achievement	Ref.
PVDF-HFP	Graphene	1 M LiPF$_6$ in EC/DMC/EMC (1:1:1)	88/470	3.61×10^{-3}/149 (C)	PVDF-HFP/graphene GPE by NIPS; Increased porosity, uptake and ionic conductivity.	[18]
PVDF-HFP	HMSS	1 M LiPF$_6$ in EC/DEC (1:1)	~70/285	2.57×10^{-3} (25 °C)/-	Development of PVDF-HFP with HMSS separators; Increased wettability and porosity.	[98]
PVDF-HFP	Li$_{1.3}$Al$_{0.3}$Ti$_{1.7}$(PO$_4$)$_3$	1 M LiTFSI + 0.25 M LiNO$_3$ in DME/DOL (1:1)	34/143.9	8.8×10^{-4} (25 °C)/1614	Ceramic/polymer membrane for lithium-sulfur cells; High ionic conductivity and discharge capacity.	[99]
PVDF-HFP	LiTSFI/SN	-	-/-	1.97×10^{-3} (20 °C)/-	Production of supercapacitors with GO electrodes and GPE; High ionic conductivity.	[100]
PVDF-HFP	LLTO	1 M LiPF$_6$ in EC/DMC/EMC (1:1:1)	69.8/497	13.897×10^{-3} (25 °C)/155.56	Incorporation of LLTO in a PVDF-HFP separator; Improved ionic conductivity.	[101]
PVDF-HFP	PI	1 M LiPF$_6$ in EC/DMC (1:1)	73/350	1.46×10^{-3}/-	Evaluation of a bicomponent electrospinning method to produce the separator. Good physical properties and improved electrochemical stability.	[102]
PVDF-HFP	PET/SiO$_2$	1 M LiPF$_6$ in EC/DEC (1:1)	60/-	9.3×10^{-4}/-	Separator with an organized porous structure, with benefits for cell operation at high C-rates; Excellent cell performance.	[103]
PVDF-HFP	MgAl$_2$O$_4$	1 M LiPF$_6$ in EC:DEC (1:1, v:v)	-/-	2.80×10^{-3}/140 (0.1C)	Influence of different quantities of the MgAl$_2$O$_4$ filler in the membrane; Good ionic conductivity.	[104]
PVDF-HFP	MgAl$_2$O$_4$	1 M LiPF$_6$ in EC/DEC (1:1, w:w)	60/81	10^{-3} (30 °C)/140 (C/10)	MgAl$_2$O$_4$ as filler of thin and flexible separator; Good thermal stability and stable cycling performance.	[105]
PVDF-HFP	Mg(OH)$_2$	1.15 M LiPF$_6$ in EC/EMC (3:7, v:v)	64/115	8.08×10^{-4}/105 (C/2)	Upgrading the battery safety operation by the addition of metal hydroxides in composite separators; High thermal stability and good capacity retention.	[106]
PVDF-HFP	MMT	1 M LiPF$_6$ in EC/DEC (1:1, v:v)	40/251	9.01×10^{-4}/105 (0.1C)	Use of montmorillonite as filler; High thermal stability and stable cycling performance.	[107]
PVDF-HFP	NaA	1 M LiPF$_6$ in EC/DEC (1:1, v:v)	65/194	2.1×10^{-3}/-	Separator with incorporation of NaA zeolite; Excellent thermal stability and wettability.	[108]

Table 5. *Cont.*

Materials	Fillers	Electrolyte Solution	Porosity and Uptake (%)	Conductivity (S·cm^{-1}) and Capacity (mAh·g^{-1})	Main Goal/Achievement	Ref.
PVDF-HFP	NaAlO$_2$	0.5 M NaTf/EMITf	-/-	5.5–6.5 × 10^{-3} (25 °C)/-	Introduction of NaAlO$_2$ in a gel polymer electrolyte; Improved ionic conductivity.	[14]
PVDF-HFP	m-SBA15	1 M LiPF$_6$ in EC/DEC (1:1)	-/82.83	3.23 × 10^{-3}/156 (0.1C)	A PVDF-HFP composite membrane with m-SBA15 as filler; High coulomb efficiency.	[109]
PVDF-HFP	m-SBA15	1 M LiPF$_6$ in EC/DEC (1:1)	-/85.36	3.78 × 10^{-3}/198.6 (0.1C)	Effect of the addition of a silica filler on a PVDF-HFP composite matrix separator; High coulomb efficiency.	[110]
PVDF-HFP	OIL	1 M LiPF$_6$ in EC/DEC (1:1)	-/13	2 × 10^{-3} (25 °C)/141 (C)	Synthesis of OIL from a phenolic epoxy resin; Non-flammability, good cell performance.	[111]
PVDF-HFP	SiO$_2$	1 M LiPF$_6$ in EC/DMC (1:2)	65.41/217	-/124.5 (C)	Synthesis of dual asymmetric structure separators with SiO$_2$ particles; High thermal stability and electrolyte uptake.	[16]
PVDF-HFP	SiO$_2$	1 M LiPF$_6$ in DMC/EMC/DC/VC (46.08:22.91:27.22:3.79)	26.7/202	8.47 × 10^{-4} (25 °C)/154.4	Composite separator with SiO$_2$; Improved thermal stability and cycling performance.	[112]
PVDF-HFP	TiO$_2$	1 M LiPF$_6$ in EC/DMC/EMC (1:1:1, v:v:v)	58/330	3.45 × 10^{-3}/122 (10C)	Evaluation of the performance of a nanocomposite polymer membrane with addition of TiO$_2$; Excellent electrochemical performance.	[85]
PVDF-HFP	ZrO$_2$	1 M LiPF$_6$ in EC/DEC (1:1)	71/182	1.48 × 10^{-3} S·cm^{-1} (25 °C)/126.8 mAhg^{-1} (0.5C)	Preparation of ZrO$_2$/PVDF-HFP by the dip-coating method. High wettability, ionic conductivity, and thermal resistance.	[113]
PVDF-HFP	ZrO$_2$	1 M LiPF$_6$ in EC/EMC (1:3)	-/-	2.06 × 10^{-3} (25 °C)/149.7	Improvement of the electrochemical properties of a electrospun membrane. High uptake and ionic conductivity.	[114]
PVDF-HFP	ZrO$_2$	1 M LiPF$_6$ in EC/DEC/DMC (1:1:1)	87.53/351.2	3.2 × 10^{-4}/646 (0.2C)	Inorganic fibers as substrates to separators; High thermal stability and good mechanical properties.	[115]
PVDF-HFP	ZrO$_2$	1 M LiPF$_6$ in EC/DMC (1:1)	60/160	10^{-3} (25 °C)/75 (C)	Development of thin and flexible ZrO$_2$ separators. High porosity and thermal stability.	[116]

Table 5. Cont.

Materials	Fillers	Electrolyte Solution	Porosity and Uptake (%)	Conductivity (S·cm^{-1}) and Capacity (mAh·g^{-1})	Main Goal/Achievement	Ref.
PVDF-HFP	ZrO$_2$	1 M LiPF$_6$ in EC/DMC (1:1)	95.7/481	2.695 × 10^{-3} (25 °C)/-	Incorporation of ZrO$_2$ in PVDF-HFP electrospun membranes; High ionic conductivity and cycling stability.	[117]
PVP/PVDF	Black carbon nanoparticles	6 M KOH	-/-	-/-	Production of separators for supercapacitor applications Improved thermal and mechanical properties.	[19]
PP/PVDF-HFP	PMMA	1 M LiPF$_6$ in EC/DMC (1:1, vv)	77.9/212	1.57 × 10^{-3}/138 (0.2C)	Physical and electrochemical performances of a PP/PVDF-HFP/PMMA composite separator; Enhanced thermal stability and electrolyte uptake.	[52]
PP/PVDF-HFP	SiO$_2$	1 M LiPF$_6$ in EC/DEC (1:1, vv)	-/290	1.76 × 10^{-3}/150 (0.2C)	PP/PVDF-HFP separator, with the inclusion of SiO$_2$ nanoparticles; Favorable chemical stability and discharge capacity.	[118]
PI/PVDF-HFP	TiO$_2$	1 M LiPF$_6$ in EC/DEC (1:1, vv)	-/-	1.88 × 10^{-3}/161 (0.5C)	Electrospun PI/PVDF-HFP membrane, with addition of TiO$_2$ nanoparticles; Excellent electrochemical properties.	[119]

Several fillers such as n-butanol [90], SiO$_2$ [103], ZnO [86] MgAl$_2$O$_4$ [105], and MMT [107] particles were used with PVDF and its copolymer composites in order to improve thermal and mechanical stability as well as the ionic conductivity value.

Mechanical improvement of separators has been achieved by developing sandwich-type composite separators, by a successive electrospinning method and based on PMIA [79].

The addition of DNA-CTMA in a PVDF matrix allows the development of flexible membranes, with interesting mechanical properties, highlighting its favorable stretch properties, allowing foldable separators with elevated elasticity [71].

The addiction of cellulose nanoparticles in the separator structure proved to significantly increase the mechanical strength of the membrane. It also improves the wettability and induces the β-phase formation in PVDF. However, the presence of NCC reduces the ionic conductivity of the membrane [73].

The use of SiO$_2$ nanoparticles in a PVDF electrospun separator can raise the mechanical strength of the membrane, thus leading to a more tough and durable battery [106].

Improved security operation for lithium ion batteries, due to suitable flammability resistance, has been addressed by developing PVDF/LiPVAOB composites membranes [33].

The direct application of a ceramic suspension of PVDF/Al$_2$O$_3$ in the electrode, resulting in a separator-cathode assembly, enhances the adhesion between these structures, and improves electrochemical cell performance [66].

PVP/PVDF membranes incorporated with carbon black nanoparticles were produced for supercapacitor applications. The separators showed improvements in mechanical properties and dielectric constant values [19].

GPEs based on boron-containing cross-linker proved to have high thermal resistance, maintaining their dimensional stability up to 150 °C, due to their stable PVDF matrix. Also, ionic conductivity and electrochemical stability were improved when compared to commercial separators [68].

Studies on the influence of solvents in nanoclay/PVDF separators showed that using DMAc as a solvent improves the porosity and electrolyte uptake of the membrane when compared with most used solvents such as NMP or DMF. Furthermore, the addition of PVP to the separator structure contributed to increase the pore size and to reduce the degree of crystallinity [72].

The addition of a metal-organic framework to a polymer structure proved to increase the conductivity of the produced membrane without needing an electrolyte. The membrane also showed high durability and good mechanical properties [76].

The dipping of PVDF nanofiber membranes into Al$_2$O$_3$ proved to improve the thermal stability of the produced separator and its ionic conductivity. It also shows a low discharge capacity decay, even at high discharge rates [111].

A double-layer separator was prepared with PVDF and reduced graphene oxide, for lithium-sulfur batteries. It is shown that the two layers combined their properties to enhance the thermal stability of the membrane and the cycling performance of the cells [82].

The use of inorganic fibers as substrate for separators lead to improved thermal and mechanical stability when compared to commercial membranes. It was also proven that it enhanced the electrochemical performance of lithium ion cells [115].

CNF/PVDF composite membranes showed greater performance when applied in Li-S batteries, with enhanced cycling stability. The produced batteries retained a capacity of 768.6 mAhg^{-1} after 200 cycles at a 0.5 C rate [20]. The development of PVDF-C separators by the phase-inversion method for Li-S batteries also leads to outstanding electrochemical performance results, associated with the presence of the conductive carbon network in the polymer matrix [69].

In the search for more environmental friendly materials, a separator with PVDF, cellulose acetate and Al(HO)$_3$ particles was developed by non-solvent induced phase separation (NIPS), the microstructure being presented in Figure 6a. This membrane exhibited high porosity, electrolyte uptake, and ionic conductivity, as well as good cycling capacity, even at high C-rates, as demonstrated in Figure 6b [70].

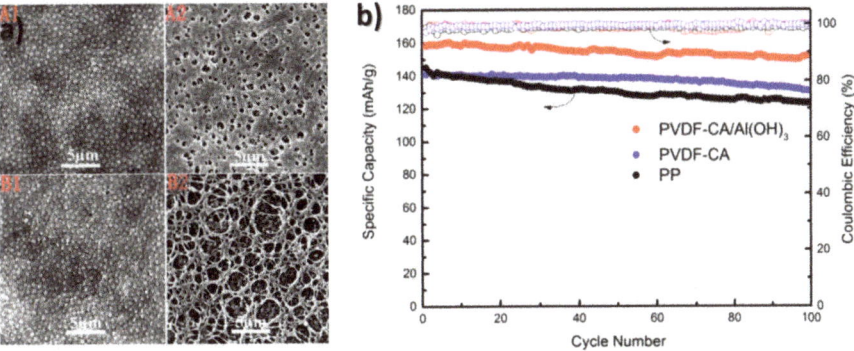

Figure 6. (a) SEM images of separators microstructure and (b) cycle performance of cells assembled [70], with copyright permission from Elsevier.

PVDF was also used in the study of the potential of zeolitic imidazolate framework-4 in separators. The prepared membranes showed high thermal stability, porosity, ionic conductivity, and cycling performance when compared with conventional separators [120].

The incorporation of Meldrum's acid groups in the PVDF structure proved to increase the ionic conductivity of the membrane, as well as the cycling performance, in particular at high C-rates [74].

PVDF/PFSA electrospun nanofibers allow the development of membrane with high mechanical stability and ionic conductivity with high discharge capacity and cycling stability [81].

A GPE membrane was developed by blending PVDF with PEO and ZrO_2. This membrane showed high electrolyte uptake, and excellent rate performance and discharge capacity for application in lithium-sulfur batteries [88].

Electrospun membranes with Octaphenyl-POSS nanoparticles showed a significant improvement in porosity and electrolyte uptake. For a ratio of 2:100 ($w:w$), the separator proved to have high mechanical stability, ionic conductivity, and thermal stability [77].

A nonaflate anion-based IL and lithium salt was introduced on a GPE, allowing the development of a membrane with high thermal stability and electrochemical properties. When used alongside with a $LiCoO_2$ cathode, this separator also showed good discharge capacity and capacity of retention [96].

The addition of $MgAl_2O_4$ as filler in electrospun fibrous PVDF-HFP separators contributes to improving electrochemical performance, with high discharge capacity and excellent cycle life results [104].

The integration of m-SBA15 as filler in a polymer matrix, on the other hand, is advantageous as it decreases the degree of crystallinity of PVDF-HFP, increasing electrolyte uptake and enhancing the ionic conductivity [109,110].

The enhancement of the electrochemical performance has been extensively addressed by composites membranes with TiO_2 nanoparticles [119], and clay nanosheets [95], the latter improving the interfacial areal connection between the polymer structure and clay, facilitating ion transport.

The NaA zeolite is considered to be a very interesting material for incorporation as filler in lithium ion battery separators. It allows the formation of voids in the composite separator structure, which are filled with electrolyte, substantially increasing the ionic conductivity [108].

The safety operation of lithium ion batteries can be upgraded by the addition of metal hydroxides such as $Al(OH)_3$ and $Mg(OH)_2$, in PVDF-HFP composite separators. These metal hydroxides endow a fire-retardant behavior to the cells, due to their natural thermal stability [91].

Kuo et al. synthesized an oligomeric ionic liquid from a phenolic epoxy resin. By blending this ionic liquid with PVDF-HFP, a high performance, non-flammable gel polymer membrane was obtained. This membrane exhibits high ionic conductivity, although with a low liquid electrolyte uptake (<50%) [111].

The addiction of ZrO_2 filler increases the porosity, ionic conductivity, and thermal resistance of the PVDF membranes. The presence of polar constituents and high connected interstitial voids facilitate electrolyte absorption, increasing the ionic conductivity and the performance of the membranes [113]. When a layer of ZrO_2 was added between two layers of PVDF-HFP, the obtained separator presented even better electrochemical properties [114].

Graphene oxide nanosheets incorporated during the phase inversion of PVDF-HFP improve electrochemical battery performances of the produced separators, as well as thermal stability and the mechanical properties of the membrane [97].

HMSS/PVDF-HFP composite separators with improved porosity were developed; the presence of SiO_2 spheres created a well-developed microporous structure, leading to higher wettability and ionic conductivity [98].

The incorporation of a superfine LLTO in a PVDF-HFP separator enhanced the ionic conductivity of the membrane. It was also been shown that a cell with a this type of separator presents improved discharge capacity and rate performance [101].

Bohemite composite separators were produced, exhibiting cycling performances comparable to the conventional ones. These membranes are also safer because of the limitation to Li dendrite formation, preventing the occurrence of short circuits [67].

A comparative study of Al_2O_3 and $NaAlO_2$ particles in a gel polymer electrolyte proved that $NaAlO_2$ membranes present higher ionic conductivity than Al_2O_3, as well as improved mechanical properties [14].

ZrO_2 membranes with PVDF-HFP as a binder were produced by solvent casting methods. These separators present high porosity and thermal stability, but show lower mechanical strength than commercially available membranes [116].

A GPE produced by thermal crosslinking of PEGDA and PEGMEA proved to be compatible with lithium ion batteries, with a high coulombic efficiency of 94% after 100 cycles [78].

Liu et al. produced a GPE with PVDF-HFP and graphene via NIPS. The addition of a small concentration of graphene (0.002 wt. %) proved to significantly improve the properties of the membrane by increasing porosity, electrolyte uptake, ionic conductivity, and cycling performance, when compared to commercial separators [18].

Regardless of the fillers type used, Table 5 shows that most of the work is devoted to increasing ionic conductivity and electrochemical performance compared to the pure matrix. In particular, inert oxide ceramics (Al_2O_3, TiO_2, SiO_2, ZrO_2) reduce the degree of crystallinity, and enhance mechanical properties and ionic conductivity value. Carbon materials (CNF, Graphene, rGO) improve safety and interfacial stability between electrodes and separator membranes, and lithium fillers such as $Li_{1.3}Al_{0.3}Ti_{1.7}(PO_4)_3$, LiTSFI, and LLTO increase ionic conductivity value of the separators.

In addition, there are other fillers types such as zeolites and clays that are being intensely used for the development of separators, allowing the improvement of electrochemical behavior.

3.4. Polymer Blend Separator Membranes

Finally, another type of separator membrane are polymer blends where two different polymers with complementary properties are used; for example one showing excellent mechanical properties and the other with a hydrophilic character. The main properties of polymer blends based on PVDF and its copolymer are presented in Table 6.

Table 6. Polymer blends based on PVDF and co-polymers with main properties, goal, and achievement.

Materials	Blends	Electrolyte Solution	Porosity and Uptake (%)	Conductivity (S·cm^{-1}) and Capacity (mAh·g^{-1})	Main Goal/Achievement	Ref
PVDF	HDPE	1 M LiPF$_6$ in EC/DEC/DMC (1:1:1)	58/260	2.54×10^{-3} S·cm^{-1} (25 °C)/156.1 mAhg^{-1} (0.1C)	Production of a sponge-like PVDF/HDPE film; High ionic conductivity and cycling performance.	[121]
PVDF	HTPB-g-MPEG	1 M LiPF$_6$ in EC/DMC/EMC (1:1:1)	56/350	3.1×10^{-3}/116 (C)	Enhancement of the stability of entrapped liquid electrolyte and corresponding ion conductivity.	[122]
PVDF	MC	1 M LiPF$_6$ in EC/DEM/EMC (1:1:1, w:w:w)	-/138.6	1.5×10^{-3}/110 (C)	PVDF composite separator with cellulose material; Excellent electrochemical performance.	[123]
PVDF	MEP	1 M TEABF$_4$ in AN/PC and 1 M LiPF$_6$ in EC/DEC	77/-	1.3×10^{-2}/-	Manufacturing by phase inversion, with MEP as a cross-linking agent; Good mechanical strength.	[34]
PVDF	NCC	1 M LiFAP in EC/DMC (1:1)	-/-	-/-	Separators with applications in hybrid electric vehicles; Favorable performance at high-voltage cells.	[124]
PVDF	NCC	1 M LiPF$_6$ in EC/DMC (1:1)	-/-	-/108 (1C)	Separators with applications in hybrid electric vehicles; Influence on high-rate cell operation.	[125]
PVDF	PAN	1 M LiPF$_6$ in EC/DMC/DEC (1:1:1)	77.7/414.5	2.9×10^{-3} (25 °C)/-	Improved thermal and mechanical properties; High cycling stability.	[126]
PVDF	PAN	1 M LiPF$_6$ in EC/DMC/EMC (1:1:1)	-/320	1.45×10^{-3}/145.71 (0.2C)	Production of an electrospun blended membrane; High thermal and mechanical stability.	[127]
PVDF	PBA	1 M LiPF$_6$ in EC/DEC/DMC (1:1:1)	-/120	8.1×10^{-4} (25 °C)/95 (0.1C)	Preparation of cross-linked PBA/PVDF GPE; Good cycling stability.	[128]
PVDF	PDMS-g-(PPO-PEO)	1 M LiPF$_6$ in EC/DMC/EMC (1:1:1, w:w:w)	80.1/512	4.5×10^{-3}/120 (1C)	Porous separator; Good electrochemical stability.	[129]
PVDF	PEGDA	1 M LiPF$_6$ in EC/DMC (1:1)	-/-	3.3×10^{-3}/117 (0.1C)	Separator produced by thermal polymerization; High capacity retention.	[130]
PVDF	PEO	1 M LiPF$_6$ in EC/DMC (1:1)	/530	-/-	Production of blended membranes by electrospinning; improved conductivity and uptake.	[131]

Table 6. Cont.

Materials	Blends	Electrolyte Solution	Porosity and Uptake (%)	Conductivity (S·cm^{-1}) and Capacity (mAh g^{-1})	Main Goal/Achievement	Ref
PVDF	PEO	1 M LiPF$_6$ in EC/DMC (1:1)	-/527	-/-	Development of electrospun membranes; High electrolyte uptake, low shutdown temperature.	[132]
PVDF	PET	-	80/270	-/-	Synthesis of a hybrid separator; High wettability and electrolyte uptake.	[133]
PVDF	PI	1 M LiPF$_6$ in EC/PC/DEC/VC (35.4:17.2:45.1:2.3)	-/-	$1.3 \times 10^{-3}/141$	Preparation of the separator by electrospinning; Improved thermal stability and mechanical properties.	[134]
PVDF	PMMA/CA	1 M LiPF$_6$ in EC/DMC (1:1, w:w)	99.1/323	-/-	Elevated porosity and electrolyte uptake.	[135]
PVDF	P(MMA-co-PEGMA)	1 M LiPF$_6$ in EC/EMC/DMC (1:1:1, w:w:w)	-/372	$3.01 \times 10^{-3}/-$	Porous separator; Improved capacity retention.	[136]
PVDF	PMMA/SiO$_2$	-	80.1/293.2	$1.97 \times 10^{-3}/-$	Evaluation of the effect of a PMMA and SiO$_2$ blend on a PVDF electrospun membrane as a separator; High electrolyte uptake and improved ionic conductivity.	[137]
PVDF	PVP	1 M Et$_4$N-BF$_4$/PC	-/360	1.8×10^{-3} (25 °C)/-	Separators for supercapacitors; High uptake and power density.	[138]
PVDF	TAIC	1 M TEABF$_4$ in AN/PC and 1 M LiPF$_6$ in EC/DEC	75/-	$1.4 \times 10^{-2}/-$	Manufacturing of separator by phase inversion, with TAIC as cross-linking agent. High ionic conductivity.	[34]
PVDF-TrFE	PEO	1 M LiTFSI in PC	44.5/107	$5.4 \times 10^{-4}/124$ (C/5)	Research on the physical and chemical properties of a PVDF-TrFE/PEO blend Favorable cycling performance.	[139]
PVDF-HFP	CA	1 M LiPF$_6$ in EC/DMC/EMC (1:1:1, v:v:v)	66.36/355	$6.16 \times 10^{-3}/138$ (0.2C)	Investigation of the use of CA from waste cigarette filters, in PVDF-HFP membranes; Good electrochemical performance and excellent thermal stability.	[140]
PVDF-HFP	HDPE	-	71/300	2.97×10^{-3} (25 °C)/140.5 (C)	Preparation of the separator by non-solvent-induced phase separation; High ionic conductivity.	[141]
PVDF-HFP	PANI	1 M LiPF$_6$ in EC/DMC (1:1)	83/270	$1.96 \times 10^{-3}/-$	High thermal stability, electrolyte uptake, and ionic conductivity	[142]

Table 6. Cont.

Materials	Blends	Electrolyte Solution	Porosity and Uptake (%)	Conductivity (S·cm^{-1}) and Capacity (mAh·g^{-1})	Main Goal/Achievement	Ref
PVDF-HFP	PEG/PEGDMA	1 M LiClO$_4$ in EC/DEC (1:1, v:v)	71/212	1.70×10^{-3}/-	Investigation into a strengthened electrospun nanofiber membrane separator; High porosity and electrolyte uptake.	[143]
PVDF-HFP	PLTB	1 M LiPF$_6$ in EC/DMC (1:1, v:v)	70/260	1.78×10^{-3}/138 (0.5C)	Excellent electrochemical performance.	[144]
PVDF-HFP	PSx-PEO3	1 M LiTFSI in EC/DMC (1:1, w:w)	-/520	4.2×10^{-4} (20 °C)/123 (C)	Production of a safe PVDF-HFP blended membrane, which can be sprayed; Elevated electrolyte uptake.	[145]
PVDF-HFP	PVSK	1 M LiTFSI + 0.25 M LiNO$_3$ in DME/DOL (1:1)	27/-	-/1220	Improved cycling performance.	[146]
PVDF-HFP	PVC	1 M LiPF$_6$ in EC/DMC (1:2)	62/230	1.58×10^{-3}/125 (0.1 C)	Tri-layer polymer membrane; Good mechanical and thermal stability.	[44]
PEI/PVDF	x-PEGDA	1 M LiPF$_6$ in EC/DMC/EMC (1:1:1)	64.6/235.6	1.38×10^{-3} (25 °C)/160.3 (0.2C)	Production of x-PEGDA-coated PEI/PVDF membranes; high wettability, porosity, and ionic conductivity.	[147]

PVDF composite separators with methyl cellulose as host of gel polymer electrolyte allows the development of low cost and environmentally friendlier separators with excellent mechanical, thermal, and electrochemical performances [123].

A trilayer porous membrane of PVDF-HFP with PVC as the middle layer was developed. It was shown that a good porosity and uptake value can be achieved, though the mechanical stability was negatively affected [44].

Cells produced with PVDF-NCC separators presented good battery performance at high C-rates, which is very critical for meeting the minimum and maximum power-assist requirements for integration in hybrid electric vehicles [124,125].

A mechanically strengthened electrospun composite PVDF-HFP/PEG/PEGDMA separator was developed. PEG and PEGDMA allow the improvement of the mechanical strength of the composite membrane, which is confirmed by the existence of physical bonded structures [143].

P(MMA-co-PEGMA) and PDMS-g-(PPO-PEO) copolymers within PVDF allow the reduction of the crystallinity of the PVDF matrix, and gently improve the electrolyte uptake, thus leading to an enhanced ionic conductivity [129,136].

PLTB can be successfully used in a PVDF-HFP composite separator. In comparison with a typical PP separator, it is more safe and efficient, due to its thermal and electrochemical stability. This separator is very promising in terms of security operation, because of flame retardant characteristics [144].

An eco-friendly technique to recover cellulose acetate from wasted cigarette filters (Figure 7) was developed, and the material can be integrated in a PVDF/CA membrane for lithium ion batteries, which presents a good performance [140].

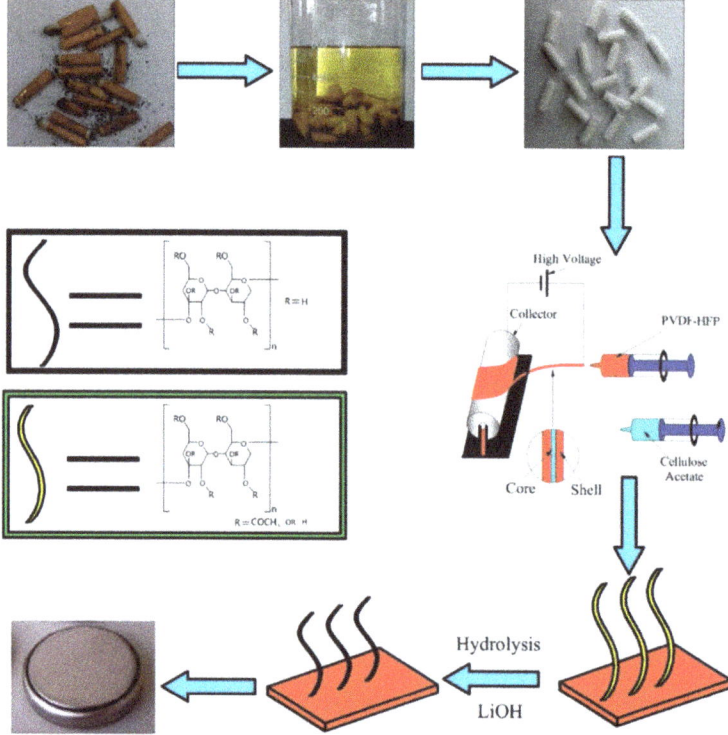

Figure 7. Preparation of PVDF-HFP/CA nanofiber separators for lithium ion batteries [140], with copyright permission from the American Chemical Society.

PVDF separators were manufactured by a phase inversion technique, with two different cross-linking agents (TAIC and MEP) and with the application of gamma radiation. The produced membranes are characterized by good mechanical behavior and low electrical resistance [34].

Electrospun PVDF membranes blended with PMMA/SiO$_2$ showed good porosity and elevated electrolyte uptake [137]. Blending with PI further enhanced their thermal and mechanical properties, ensuring a better battery performance than commercial PE separators [134].

PVDF/PEO blend membranes show an increase of the ionic conductivity and electrolyte uptake when compared with PVDF membranes. The improved wettability and porosity in x-PEGDA-coated PEI/PVDF membranes has been also reported [147].

PVDF-HFP/HDPE membranes were prepared by non-solvent induced phase separation. This separator presents good cycling performance in lithium ion batteries and a high ionic conductivity [141]. Further studies showed an increased discharge capacity of these membranes, by decreasing the size of the HDPE fillers [121].

PVDF/PAN blend separators were produced by TIPS [126] and electrospinning [127], with improved thermal and mechanical properties. The best PVDF/PAN ratio was 90:10. Despite the lower ionic conductivity when compared with conventional separators, these membranes showed higher cycle and C-rate performance [126].

PVDF/PAN electrospun membrane have excellent dimensional stability even at high temperatures, high electrolyte uptake and ionic conductivity, and superior discharge capacity [127].

The blending of PVDF and PEO in an electrospun membrane proved to increase significantly the electrolyte uptake of the separator, while decreasing the shutdown temperature [132]

Cross-linked PBA/PVDF GPE were prepared by soaking semi-interpenetrating polymer networks with liquid electrolyte. For a PBA/PVDF ratio of 1:0.5, the best results of electrolyte uptake, ionic conductivity, and cycling stability were obtained [128].

A PVDF/PET hybrid separator was produced via a mechanical pressing process. The obtained membrane presented high wettability and electrolyte uptake, while maintaining good thermal stability [133].

The introduction of PANI in a PVDF separator by the breath figure method proved to increase the electrolyte uptake and ionic conductivity of the membrane. The best results were obtained for 30% of PANI, with a uniform pore structure and excellent thermal stability [142].

The use of PVDF-HFP/PVSK membranes in lithium-sulfur batteries has been reported. It has been proved that even small amounts of PVSK (5 wt. %) increase the discharge capacity of the cell and reduce the capacity decay [146].

An increase of the use of natural polymers and biopolymers is observed for the preparation of PVDF and copolymer blends, considering the environmental issues. It is demonstrated in Table 6 that they allow to improve mechanical properties and wettability, and consequently the battery performance. In addition, the use of conductive polymers such as PANI in polymer blends has acquired special attention in recent years, considering that the electrical properties are improved without mechanical deterioration. Typically, the most commonly used PVDF and PVDF-HFP blends are developed with PAN and PEO polymers, allowing the improve thermal and mechanical stability, as well as wettability and ionic conductivity value, respectively.

4. Conclusions and Future Trends

In this review, the latest advances in PVDF-based battery separators for lithium-ion battery applications are presented.

Considering the excellent properties of PVDF and its copolymers as a separation membrane and the importance of the role of the battery separator in battery applications, this review was divided into four different sections—single polymers, surface modification, polymer composites, and blends, where, for each category, the improvement of the main properties of the separators' degree of porosity, uptake value, mechanical and thermal properties, ionic conductivity, and cycling performance, as well as safety and environmental impacts,- for the different developed materials was presented.

In the single polymer category, PVDF and PVDF-HFP stand out as the most commonly applied polymers produced by various processing techniques, with TIPS and electrospinning methods being the most commonly used to tailor microstructure (degree of porosity and pore size) to improve battery performance.

The number of research papers on surface modifications of the membranes has increased in recent years, as the surface of the polymer membrane strongly affects the uptake process. Surface modification is accomplished by coating hydrophilic polymers or plasma treatment to increase the interaction between the polymer membrane and the electrolytic solution.

Generally, the addition of fillers increases battery performance through the improvement of ionic conductivity in polymer composites, but has not yet demonstrated the best filler for PVDF and its copolymer membranes. The most commonly used fillers are inert oxide ceramics, carbon materials, and lithium fillers. The most improved properties are mechanical properties, interfacial stability, between electrodes and separator membranes and ionic conductivity value, respectively.

In relation to the polymer blends, the appearance of new blends based on natural and conductive polymers within PVDF for battery separators has been observed.

The blends of PVDF and its copolymers widely used are with PAN and PEO polymers, allowing the improvement of mechanical properties and wettability and electric properties, respectively.

The future trends for single polymer separators are to obtain single polymers with a porosity above 50% but a smaller pore size below 500 nm to prevent dendrite growth. Further, it is expected an increase in the use of ionic liquids as electrolytic solution. In relation to surface modifications, the use of poly (ionic liquids) and natural polymers as a surface modification coating of PVDF polymer membranes will be interesting, considering environmental issues.

With respect to polymer composites, future perspectives are related to improving the interaction between polymer matrix and fillers, in order to optimize filler content without decreasing electrical properties or hindering mechanical stability. Also, the use of more than one filler with complementary properties may be the way for improving cycling performance.

The progress with respect to polymer blends is related to the scalability of the fabrication process, increasing the interaction and compatibilization of the two polymers.

In summary, PVDF-based battery separators allow the tailoring of all the properties/characteristics required for a new generation of separator membranes for lithium-ion batteries with high power and excellent cycling performance.

Funding: Portuguese Foundation for Science and Technology (FCT): UID/FIS/04650/2013, PTDC/CTM-ENE/5387/2014, UID/CTM/50025/2013, project NO. 28157/02/SAICT/2017 and grants SFRH/BPD/112547/2015 (C.M.C.), including FEDER funds through the COMPETE 2020 programme and National Funds through FCT. Financial support from the Basque Government Industry Department under the ELKARTEK and HAZITEK programs is also acknowledged.

Conflicts of Interest: The authors declare no conflict of interest.

List of Symbols and Abbreviations

$(C_2H_5)_3CH_3NBF_4$	Triethylmethylammonium tetrafluoroborate
[C2mim][NTf2]	1-ethyl-3-methylimidazolium bis(trifluoromethylsulfonyl)imide
$Al(OH)_3$	Aluminum hydroxide
Al_2O_3	Aluminum oxide
AlO(OH)	Bohemite
AN	Acetonitrile
BC	Boron-containing cross-linker
CA	Cellulose acetate
CMC	carboxymethyl cellulose
CNF	Carbon nanofiber
DEC	Diethyl carbonate

DEM	Diethoxymethane
DMAc	Dimethyl acetamide
DMC	Dimethyl carbonate
DME	1,2-dimethoxyethane
DMF	Dimethyl formamide
DMSO	Dimethyl sulfoxide
DNA-CTMA	Deoxyribonucleic acid-cetyltrimethylammonium
DOL	1,3-dioxolane
EC	Ethylene carbonate
EMC	Ethyl methyl carbonate
EMImNfO-LiNfO	1-ethyl-3- methylimidazolium nonafluoro-1-butanesulfonate/lithium nonafluoro-1-butanesulfonate
EMITf	1-ethyl 3-methyl imidazolium trifluoromethane sulfonate
EMITFSI	1-ethyl-3-methyl-imidazolium bis(trifluoromethanesulfonyl)imide
EP	Ethyl propionate
$Et_4N\text{-}BF_4$	Tetraethylammonium tetrafluoroboratein
GF	Glass fiber
GO	Graphene oxides
GPE	Gel polymer electrolyte
H_2SO_4	Sulfuric acid
HDPE	High density polyethylene
HEC	Hydroxyethyl cellulose
HMSS	Hollow mesoporous silica spheres
HTPB-g-MPEG	Hydroxyl-terminated polybutadiene-grafted-methoxyl polyethylene glycol
KOH	Potassium hydroxide
$LiClO_4$	Lithium percholorate
$LiCoO_2$	Lithium cobalt oxide
LiFAP	Lithium Tris(pentafluoroethane)-trifluorophosphate
LiNfO/BMImNfO	Lithium nonafluorobutanesulfonate/1-butyl-3-me-thylimidazolium nonafluorobutanesulfonate
$LiNO_3$	Lithium nitrate
$LiPF_6$	Lithium hexafluorophosphate
LiPVAOB	Lithium polyvinyl alcohol oxalate borate
Li-S	Lithium-sulfur
LiTFSI	lithium bis(trifluoromethanesulfonyl)imide
LLTO	Li0.33La0.557TiO3
MA	Meldrum's acid
MC	Methyl cellulose
MEP	Ethylene oxide-propylene oxide
$Mg(OH)_2$	Magnesium hydroxide
$MgAl_2O_4$	Magnesium aluminate
MMT	Montmorillonite
MOF-808	Zirconium (IV) metal-organic framework
m-SBA 15	Mesoporous silica
NaA	NaA zeolite
$NaClO_4$	Sodium perchlorate
NaTf	Sodium trifluoromethane sulfonate
NCC	Nanocrystalline cellulose
NIPS	Non-solvent induced phase separation
NMP	N-methyl-2-pyrrolidone
OIL	Oligomeric ionic liquid (bromide bis(tri-fluoromethane)sulfonimide)
P(MMA-co-PEGMA)	Poly(methyl methacrylate-co-poly(ethylene glycol) methacrylate)
PAN	Polyacrylonitrile
PANI	Polyaniline
PBA	Poly(butyl acrylate)
PC	Propylene carbonate
PDA	Polydopamine
PDMS-g-(PPO-PEO)	Poly(dimethylsiloxane)-graft-poly(propylene oxide)-block-poly(ethylene oxide)
PE	Polyethylene

PEG	Polyethylene glycol
PEGDA	Poly(ethylene glicol)diacrylate
PEGDMA	Polyethylene glycol dimethacrylate
PEGMEA	Poly(ethylene glycol) methyl ether acrylate
PEI	Polyetherimide
PEO	Polyethilene oxide
PET	Polyethylene terephthalate
PFSA	Perflourosulfonic acid
PI	Polyimide
PLTB	Polimeric lithium tartaric acid borate
PMIA	Poly(m-phenylene isophthalamide)
PMMA	Polymethyl methacrylate
POSS	Polyhedral oligomeric silsesquioxane
PP	Polypropylene
P-PAEK	Phenolphthaleyne-poly(aryl ether ketone)
PSx-PEO3	Polysiloxane-comb-propyl(triethylene oxide)
PSU	Poly(sulfone)
PTFE	Poly(tetrafluoroethylene)
PVA	Polyvinyl alcohol
PVC	Poly(vinyl chloride)
PVDF	Poly(vinylidene fluoride)
PVDF-co-CTFE	Polyvinylidene fluoride-co-chlorotrifluoroethylene
PVDF-co-HFP	Poly(vinylidene fluoride-co-hexafluoropropylene)
PVDF-HFP	Poly(vinylidene fluoride-co-hexafluoropropene)Poly(vinylidene fluoride-hexafluoropropylene)
PVDF-PE	Polyvinylidene difluoride-coated polyethylene
PVDF-TrFE	Poly(vinylidene fluoride-trifluoroethylene)
PVP	Polyvinylpyrrolidone
PVSK	Polyvinylsulfate potassium salt
rGO	Reduced graphene oxide
SCPC	Self-charging power cell
SiO_2	Silicon dioxide
SN	Succinonitrile
SnO_2	Tin oxide
TAIC	Triallyl isocyanurate
$TEABF_4$	Tetraethyl ammonium tetrafluoroborate
TiO_2	Titanium dioxide
TIPS	Thermal-induced phase separation
TTT	1,3,5-trially-1,3,5-triazine-2,4,6(1 H,3 H,5 H)-trione
VC	Vinylene carbonate
x-PEGDA	x-polyethylene glycol diacrylate
ZnO	Zinc oxide
ZrO2	Zirconium dioxide

References

1. Megahed, S.; Ebner, W. Lithium-ion battery for electronic applications. *J. Power Sources* **1995**, *54*, 155–162. [CrossRef]
2. Oliveira, J.; Correia, V.; Castro, H.; Martins, P.; Lanceros-Mendez, S. Polymer-based smart materials by printing technologies: Improving application and integration. *Add. Manuf.* **2018**, *21*, 269–283. [CrossRef]
3. Goodenough, J.B.; Park, K.S. The li-ion rechargeable battery: A perspective. *JACS* **2013**, *135*, 1167–1176. [CrossRef] [PubMed]
4. Lanceros-Méndez, S.; Costa, C.M. *Printed Batteries: Materials, Technologies and Applications*; Wiley: Hoboken, NJ, USA, 2018.
5. Manthiram, A. An outlook on lithium ion battery technology. *ACS Cent. Sci.* **2017**, *3*, 1063–1069. [CrossRef] [PubMed]
6. Research, A.M. *Lithium-Ion Battery Market—Global Opportunity Analysis and Industry Forecast*; Allied Market Research: London, UK, 2016; pp. 2015–2022.

7. Yoshio, M.; Brodd, R.J.; Kozawa, A. *Lithium-Ion Batteries Science and Technologies*; Springer: Berlin, Germany, 2009.
8. Wu, Y.P. *Lithium-Ion Batteries Fundamentals and Applications*; CRC Press: Boca Raton, FL, USA, 2015.
9. Costa, C.M.; Rodrigues, H.M.; Gören, A.; Machado, A.V.; Silva, M.M.; Lanceros-Méndez, S. Preparation of poly(vinylidene fluoride) lithium-ion battery separators and their compatibilization with ionic liquid—A green solvent approach. *Chem. Select* **2017**, *2*, 5394–5402. [CrossRef]
10. Arora, P.; Zhang, Z.J. Battery separators. *Chem. Rev.* **2004**, *104*, 4419–4462. [CrossRef] [PubMed]
11. Costa, C.M.; Silva, M.M.; Lanceros-Méndez, S. Battery separators based on vinylidene fluoride (vdf) polymers and copolymers for lithium ion battery applications. *RSC Adv.* **2013**, *3*, 11404–11417. [CrossRef]
12. Shi, C.; Dai, J.; Huang, S.; Li, C.; Shen, X.; Zhang, P.; Wu, D.; Sun, D.; Zhao, J. A simple method to prepare a polydopamine modified core-shell structure composite separator for application in high-safety lithium-ion batteries. *J. Membr. Sci.* **2016**, *518*, 168–177. [CrossRef]
13. Nho, Y.C.; Sohn, J.Y.; Shin, J.; Park, J.S.; Lim, Y.M.; Kang, P.H. Preparation of nanocomposite γ-al$_2$o$_3$/polyethylene separator crosslinked by electron beam irradiation for lithium secondary battery. *Radiat. Phys. Chem.* **2017**, *132*, 65–70. [CrossRef]
14. Hashmi, S.A.; Bhat, M.Y.; Singh, M.K.; Sundaram, N.T.K.; Raghupathy, B.P.C.; Tanaka, H. Ionic liquid-based sodium ion-conducting composite gel polymer electrolytes: Effect of active and passive fillers. *J. Solid State Electrochem.* **2016**, *20*, 2817–2826. [CrossRef]
15. Wu, D.; Deng, L.; Sun, Y.; Teh, K.S.; Shi, C.; Tan, Q.; Zhao, J.; Sun, D.; Lin, L. A high-safety pvdf/Al$_2$O$_3$ composite separator for Li-ion batteries via tip-induced electrospinning and dip-coating. *RSC Adv.* **2017**, *7*, 24410–24416. [CrossRef]
16. Wang, Y.; Zhu, S.; Sun, D.; Jin, Y. Preparation and evaluation of a separator with an asymmetric structure for lithium-ion batteries. *RSC Adv.* **2016**, *6*, 105461–105468. [CrossRef]
17. Wang, H.; Zhang, Y.; Gao, H.; Jin, X.; Xie, X. Composite melt-blown nonwoven fabrics with large pore size as li-ion battery separator. *Int. J. Hydrog. Energy* **2016**, *41*, 324–330. [CrossRef]
18. Liu, J.; Wu, X.; He, J.; Li, J.; Lai, Y. Preparation and performance of a novel gel polymer electrolyte based on poly(vinylidene fluoride)/graphene separator for lithium ion battery. *Electrochim. Acta* **2017**, *235*, 500–507. [CrossRef]
19. Jabbarnia, A.; Khan, W.S.; Ghazinezami, A.; Asmatulu, R. Investigating the thermal, mechanical, and electrochemical properties of pvdf/pvp nanofibrous membranes for supercapacitor applications. *J. Appl. Polym. Sci.* **2016**, *133*, 1–10. [CrossRef]
20. Wang, Z.; Zhang, J.; Yang, Y.; Yue, X.; Hao, X.; Sun, W.; Rooney, D.; Sun, K. Flexible carbon nanofiber/polyvinylidene fluoride composite membranes as interlayers in high-performance lithium[sbnd]sulfur batteries. *J. Power Sources* **2016**, *329*, 305–313. [CrossRef]
21. Ye, Y.-S.; Rick, J.; Hwang, B.-J. Ionic liquid polymer electrolytes. *J. Mater. Chem. A* **2013**, *1*, 2719–2743. [CrossRef]
22. Kim, J.F.; Jung, J.T.; Wang, H.H.; Lee, S.Y.; Moore, T.; Sanguineti, A.; Drioli, E.; Lee, Y.M. Microporous pvdf membranes via thermally induced phase separation (tips) and stretching methods. *J. Membr. Sci.* **2016**, *509*, 94–104. [CrossRef]
23. Ribeiro, C.; Costa, C.M.; Correia, D.M.; Nunes-Pereira, J.; Oliveira, J.; Martins, P.; Gonçalves, R.; Cardoso, V.F.; Lanceros-Méndez, S. Electroactive poly(vinylidene fluoride)-based structures for advanced applications. *Nat. Protoc.* **2018**, *13*, 681. [CrossRef] [PubMed]
24. Martins, P.; Lopes, A.C.; Lanceros-Mendez, S. Electroactive phases of poly(vinylidene fluoride): Determination, processing and applications. *Prog. Polym. Sci.* **2014**, *39*, 683–706. [CrossRef]
25. Nalwa, H.S. *Ferroelectric Polymers: Chemistry: Physics, and Applications*; CRC Press: Boca Raton, FL, USA, 1995.
26. Sousa, R.E.; Ferreira, J.C.C.; Costa, C.M.; Machado, A.V.; Silva, M.M.; Lanceros-Mendez, S. Tailoring poly(vinylidene fluoride-co-chlorotrifluoroethylene) microstructure and physicochemical properties by exploring its binary phase diagram with dimethylformamide. *J. Polym. Sci. Part B Polym. Phys.* **2015**, *53*, 761–773. [CrossRef]
27. Sousa, R.E.; Nunes-Pereira, J.; Ferreira, J.C.C.; Costa, C.M.; Machado, A.V.; Silva, M.M.; Lanceros-Mendez, S. Microstructural variations of poly(vinylidene fluoride co-hexafluoropropylene) and their influence on the thermal, dielectric and piezoelectric properties. *Polym. Test.* **2014**, *40*, 245–255. [CrossRef]

28. Costa, C.M.; Rodrigues, L.C.; Sencadas, V.; Silva, M.M.; Rocha, J.G.; Lanceros-Méndez, S. Effect of degree of porosity on the properties of poly(vinylidene fluoride–trifluorethylene) for li-ion battery separators. *J. Membr. Sci.* **2012**, *407–408*, 193–201. [CrossRef]
29. Idris, N.H.; Rahman, M.M.; Wang, J.-Z.; Liu, H.-K. Microporous gel polymer electrolytes for lithium rechargeable battery application. *J. Power Sources* **2012**, *201*, 294–300. [CrossRef]
30. Tõnurist, K.; Vaas, I.; Thomberg, T.; Jänes, A.; Kurig, H.; Romann, T.; Lust, E. Application of multistep electrospinning method for preparation of electrical double-layer capacitor half-cells. *Electrochim. Acta* **2014**, *119*, 72–77. [CrossRef]
31. Cao, C.; Tan, L.; Liu, W.; Ma, J.; Li, L. Polydopamine coated electrospun poly(vinyldiene fluoride) nanofibrous membrane as separator for lithium-ion batteries. *J. Power Sources* **2014**, *248*, 224–229. [CrossRef]
32. Saito, Y.; Morimura, W.; Kuratani, R.; Nishikawa, S. Ion transport in separator membranes of lithium secondary batteries. *J. Phys. Chem. C* **2015**, *119*, 4702–4708. [CrossRef]
33. Zhu, Y.; Xiao, S.; Shi, Y.; Yang, Y.; Hou, Y.; Wu, Y. A composite gel polymer electrolyte with high performance based on poly(vinylidene fluoride) and polyborate for lithium ion batteries. *Adv. Energy Mater.* **2014**, *4*, 1300647. [CrossRef]
34. Karabelli, D.; Leprêtre, J.C.; Dumas, L.; Rouif, S.; Portinha, D.; Fleury, E.; Sanchez, J.Y. Crosslinking of poly(vinylene fluoride) separators by gamma-irradiation for electrochemical high power charge applications. *Electrochim. Acta* **2015**, *169*, 32–36. [CrossRef]
35. Musil, M.; Pléha, D. Nonwoven separators fabrication and analysis methods. *ECS Trans.* **2015**, *70*, 127–133. [CrossRef]
36. Liang, H.Q.; Wan, L.S.; Xu, Z.K. Poly(vinylidene fluoride) separators with dual-asymmetric structure for high-performance lithium ion batteries. *Chin. J. Polym. Sci. (Engl. Ed.)* **2016**, *34*, 1423–1435. [CrossRef]
37. He, H.; Fu, Y.; Zhao, T.; Gao, X.; Xing, L.; Zhang, Y.; Xue, X. All-solid-state flexible self-charging power cell basing on piezo-electrolyte for harvesting/storing body-motion energy and powering wearable electronics. *Nano Energy* **2017**, *39*, 590–600. [CrossRef]
38. Song, J.; Liu, L.; Yang, Q.; Liu, J.; Yu, T.; Yang, F.; Crittenden, J. Pvdf layer as a separator on the solution-side of air-cathodes: The electricity generation, fouling and regeneration. *RSC Adv.* **2015**, *5*, 52361–52368. [CrossRef]
39. Song, R.; Jin, H.; Li, X.; Fei, L.; Zhao, Y.; Huang, H.; Lai-Wa Chan, H.; Wang, Y.; Chai, Y. A rectification-free piezo-supercapacitor with a polyvinylidene fluoride separator and functionalized carbon cloth electrodes. *J. Mater. Chem. A* **2015**, *3*, 14963–14970. [CrossRef]
40. Janakiraman, S.; Surendran, A.; Ghosh, S.; Anandhan, S.; Venimadhav, A. Electroactive poly(vinylidene fluoride) fluoride separator for sodium ion battery with high coulombic efficiency. *Solid State Ionics* **2016**, *292*, 130–135. [CrossRef]
41. Kundu, M.; Costa, C.M.; Dias, J.; Maceiras, A.; Vilas, J.L.; Lanceros-Méndez, S. On the relevance of the polar β-phase of poly(vinylidene fluoride) for high performance lithium-ion battery separators. *J. Phys. Chem. C* **2017**, *121*, 26216–26225. [CrossRef]
42. Jeschke, S.; Mutke, M.; Jiang, Z.; Alt, B.; Wiemhofer, H.D. Study of carbamate-modified disiloxane in porous pvdf-hfp membranes: New electrolytes/separators for lithium-ion batteries. *Chemphyschem* **2014**, *15*, 1761–1771. [CrossRef] [PubMed]
43. Karuppasamy, K.; Reddy, P.A.; Srinivas, G.; Tewari, A.; Sharma, R.; Shajan, X.S.; Gupta, D. Electrochemical and cycling performances of novel nonafluorobutanesulfonate (nonaflate) ionic liquid based ternary gel polymer electrolyte membranes for rechargeable lithium ion batteries. *J. Membr. Sci.* **2016**, *514*, 350–357. [CrossRef]
44. Angulakshmi, N.; Stephan, A.M. Electrospun trilayer polymeric membranes as separator for lithium–ion batteries. *Electrochim. Acta* **2014**, *127*, 167–172. [CrossRef]
45. Zhang, J.; Sun, B.; Huang, X.; Chen, S.; Wang, G. Honeycomb-like porous gel polymer electrolyte membrane for lithium ion batteries with enhanced safety. *Sci. Rep.* **2014**, *4*, 6007. [CrossRef] [PubMed]
46. Heo, J.; Choi, Y.; Chung, K.Y.; Park, J.H. Controlled pore evolution during phase inversion from the combinatorial non-solvent approach: Application to battery separators. *J. Mater. Chem. A* **2016**, *4*, 9496–9501. [CrossRef]
47. Lee, H.; Alcoutlabi, M.; Toprakci, O.; Xu, G.; Watson, J.V.; Zhang, X. Preparation and characterization of electrospun nanofiber-coated membrane separators for lithium-ion batteries. *J. Solid State Electrochem.* **2014**, *18*, 2451–2458. [CrossRef]

48. Laurita, R.; Zaccaria, M.; Gherardi, M.; Fabiani, D.; Merlettini, A.; Pollicino, A.; Focarete, M.L.; Colombo, V. Plasma processing of electrospun li-ion battery separators to improve electrolyte uptake. *Plasma Process. Polym.* **2016**, *13*, 124–133. [CrossRef]
49. Fang, L.-F.; Shi, J.; Li, H.; Zhu, B.K.; Zhu, L.P. Construction of porous pvdf coating layer and electrochemical performances of the corresponding modified polyethylene separators for lithium ion batteries. *J. Appl. Polym. Sci.* **2014**, *131*, 1–9. [CrossRef]
50. Kim, C.-S.; Jeong, K.M.; Kim, K.; Yi, C.-W. Effects of capacity ratios between anode and cathode on electrochemical properties for lithium polymer batteries. *Electrochim. Acta* **2015**, *155*, 431–436. [CrossRef]
51. Xu, R.; Huang, X.; Lin, X.; Cao, J.; Yang, J.; Lei, C. The functional aqueous slurry coated separator using polyvinylidene fluoride powder particles for lithium-ion batteries. *J. Electroanal. Chem.* **2017**, *786*, 77–85. [CrossRef]
52. Wu, D.; Huang, S.; Xu, Z.; Xiao, Z.; Shi, C.; Zhao, J.; Zhu, R.; Sun, D.; Lin, L. Polyethylene terephthalate/poly(vinylidene fluoride) composite separator for li-ion battery. *J. Phys. D Appl. Phys.* **2015**, *48*, 245304. [CrossRef]
53. Zhang, M.Y.; Li, M.X.; Chang, Z.; Wang, Y.F.; Gao, J.; Zhu, Y.S.; Wu, Y.P.; Huang, W. A sandwich pvdf/hec/pvdf gel polymer electrolyte for lithium ion battery. *Electrochim. Acta* **2017**, *245*, 752–759. [CrossRef]
54. Yang, W.; Yang, W.; Feng, J.; Ma, Z.; Shao, G. High capacity and cycle stability rechargeable lithium-sulfur batteries by sandwiched gel polymer electrolyte. *Electrochim. Acta* **2016**, *210*, 71–78. [CrossRef]
55. An, M.-Y.; Kim, H.-T.; Chang, D.-R. Multilayered separator based on porous polyethylene layer, Al_2O_3 layer, and electro-spun pvdf nanofiber layer for lithium batteries. *J. Solid State Electrochem.* **2014**, *18*, 1807–1814. [CrossRef]
56. Wu, D.; Shi, C.; Huang, S.; Qiu, X.; Wang, H.; Zhan, Z.; Zhang, P.; Zhao, J.; Sun, D.; Lin, L. Electrospun nanofibers for sandwiched polyimide/poly (vinylidene fluoride)/polyimide separators with the thermal shutdown function. *Electrochim. Acta* **2015**, *176*, 727–734. [CrossRef]
57. Kim, J.I.; Choi, Y.; Chung, K.Y.; Park, J.H. A structurable gel-polymer electrolyte for sodium ion batteries. *Adv. Funct. Mater.* **2017**, *27*, 1–7. [CrossRef]
58. Li, W.; Li, X.; Xie, X.; Yuan, A.; Xia, B. Effect of drying temperature on a thin pvdf-hfp/pet composite nonwoven separator for lithium-ion batteries. *Ionics* **2017**, *23*, 929–935. [CrossRef]
59. Liu, H.; Dai, Z.; Xu, J.; Guo, B.; He, X. Effect of silica nanoparticles/poly(vinylidene fluoride-hexafluoropropylene) coated layers on the performance of polypropylene separator for lithium-ion batteries. *J. Energy Chem.* **2014**, *23*, 582–586. [CrossRef]
60. Zhang, J.; Chen, S.; Xie, X.; Kretschmer, K.; Huang, X.; Sun, B.; Wang, G. Porous poly(vinylidene fluoride-co-hexafluoropropylene) polymer membrane with sandwich-like architecture for highly safe lithium ion batteries. *J. Membr. Sci.* **2014**, *472*, 133–140. [CrossRef]
61. Wu, Y.S.; Yang, C.C.; Luo, S.P.; Chen, Y.L.; Wei, C.N.; Lue, S.J. Pvdf-hfp/pet/pvdf-hfp composite membrane for lithium-ion power batteries. *Int. J. Hydrog. Energy* **2017**, *42*, 6862–6875. [CrossRef]
62. Lee, Y.; Lee, H.; Lee, T.; Ryou, M.-H.; Lee, Y.M. Synergistic thermal stabilization of ceramic/co-polyimide coated polypropylene separators for lithium-ion batteries. *J. Power Sources* **2015**, *294*, 537–544. [CrossRef]
63. Alcoutlabi, M.; Lee, H.; Zhang, X. Nanofiber-based membrane separators for lithium-ion batteries. *MRS Proc.* **2015**, *1718*. [CrossRef]
64. Kim, K.J.; Kwon, H.K.; Park, M.S.; Yim, T.; Yu, J.S.; Kim, Y.J. Ceramic composite separators coated with moisturized zro_2 nanoparticles for improving the electrochemical performance and thermal stability of lithium ion batteries. *Phys. Chem. Chem. Phys.* **2014**, *16*, 9337–9343. [CrossRef] [PubMed]
65. Shen, X.; Li, C.; Shi, C.; Yang, C.; Deng, L.; Zhang, W.; Peng, L.; Dai, J.; Wu, D.; Zhang, P.; et al. Core-shell structured ceramic nonwoven separators by atomic layer deposition for safe lithium-ion batteries. *Appl. Surf. Sci.* **2018**, *441*, 165–173. [CrossRef]
66. Xiao, W.; Zhao, L.; Gong, Y.; Wang, S.; Liu, J.; Yan, C. Preparation of high performance lithium-ion batteries with a separator–cathode assembly. *RSC Adv.* **2015**, *5*, 34184–34190. [CrossRef]
67. Holtmann, J.; Schäfer, M.; Niemöller, A.; Winter, M.; Lex-Balducci, A.; Obeidi, S. Boehmite-based ceramic separator for lithium-ion batteries. *J. Appl. Electrochem.* **2016**, *46*, 69–76. [CrossRef]

68. Shim, J.; Lee, J.S.; Lee, J.H.; Kim, H.J.; Lee, J.-C. Gel polymer electrolytes containing anion-trapping boron moieties for lithium-ion battery applications. *ACS Appl. Mater. Interfaces* **2016**, *8*, 27740–27752. [CrossRef] [PubMed]
69. Wei, H.; Ma, J.; Li, B.; Zuo, Y.; Xia, D. Enhanced cycle performance of lithium-sulfur batteries using a separator modified with a pvdf-c layer. *ACS Appl. Mater. Interfaces* **2014**, *6*, 20276–20281. [CrossRef] [PubMed]
70. Cui, J.; Liu, J.; He, C.; Li, J.; Wu, X. Composite of polyvinylidene fluoride–cellulose acetate with Al(OH)$_3$ as a separator for high-performance lithium ion battery. *J. Membr. Sci.* **2017**, *541*, 661–667. [CrossRef]
71. Kobayashi, N.; Ouchen, F.; Rau, I.; Kumar, J.; Ouchen, F.; Smarra, D.A.; Subramanyam, G.; Grote, J.G. DNA based electrolyte/separator for lithium battery application. In *Nanobiosystems: Processing, Characterization, and Applications VIII*; Society of Photo-optical Instrumentation Engineers: Bellingham, WA, USA, 2015; Volume 9557, p. 95570A.
72. Rahmawati, S.A.; Sulistyaningsih; Putro, A.Z.A.; Widyanto, N.F.; Jumari, A.; Purwanto, A.; Dyartanti, E.R. Preparation and characterization of nanocomposite polymer electrolytes poly(vinylidone fluoride)/nanoclay. In *AIP Conference Proceedings*; American Institute of Physics: College Park, MD, USA, 2016; Volume 1710, p. 030053.
73. Bolloli, M.; Antonelli, C.; Molméret, Y.; Alloin, F.; Iojoiu, C.; Sanchez, J.Y. Nanocomposite poly(vynilidene fluoride)/nanocrystalline cellulose porous membranes as separators for lithium-ion batteries. *Electrochim. Acta* **2016**, *214*, 38–48. [CrossRef]
74. Lee, Y.Y.; Liu, Y.L. Crosslinked electrospun poly(vinylidene difluoride) fiber mat as a matrix of gel polymer electrolyte for fast-charging lithium-ion battery. *Electrochim. Acta* **2017**, *258*, 1329–1335. [CrossRef]
75. Fang, C.; Yang, S.; Zhao, X.; Du, P.; Xiong, J. Electrospun montmorillonite modified poly(vinylidene fluoride) nanocomposite separators for lithium-ion batteries. *Mater. Res. Bull.* **2016**, *79*, 1–7. [CrossRef]
76. Luo, H.B.; Wang, M.; Liu, S.X.; Xue, C.; Tian, Z.F.; Zou, Y.; Ren, X.M. Proton conductance of a superior water-stable metal-organic framework and its composite membrane with poly(vinylidene fluoride). *Inorg. Chem.* **2017**, *56*, 4169–4175. [CrossRef] [PubMed]
77. Chen, H.L.; Jiao, X.N. Preparation and characterization of polyvinylidene fluoride/octaphenyl-polyhedral oligomeric silsesquioxane hybrid lithium-ion battery separators by electrospinning. *Solid State Ionics* **2017**, *310*, 134–142. [CrossRef]
78. Li, H.; Chao, C.Y.; Han, P.L.; Yan, X.R.; Zhang, H.H. Preparation and properties of gel-filled pvdf separators for lithium ion cells. *J. Appl. Polym. Sci.* **2017**, *134*, 6–11. [CrossRef]
79. Zhai, Y.; Wang, N.; Mao, X.; Si, Y.; Yu, J.; Al-Deyab, S.S.; El-Newehy, M.; Ding, B. Sandwich-structured pvdf/pmia/pvdf nanofibrous separators with robust mechanical strength and thermal stability for lithium ion batteries. *J. Mater. Chem. A* **2014**, *2*, 14511–14518. [CrossRef]
80. Xie, M.; Yin, M.; Nie, G.; Wang, J.; Wang, C.; Chao, D.; Liu, X. Poly(aryl ether ketone) composite membrane as a high-performance lithium-ion batteries separator. *J. Polym. Sci. Part A Polym. Chem.* **2016**, *54*, 2714–2721. [CrossRef]
81. Meng-Nan, H.; Jiang, Z.-Q.; Li, F.-B.; Yang, H.; Xu, Z.-L. Preparation and characterization of pfsa-pvdf blend nanofiber membrane and its preliminary application investigation. *New J. Chem.* **2017**, *41*, 7544–7552.
82. Zhu, P.; Zhu, J.; Zang, J.; Chen, C.; Lu, Y.; Jiang, M.; Yan, C.; Dirican, M.; Kalai Selvan, R.; Zhang, X. A novel bi-functional double-layer rgo–pvdf/pvdf composite nanofiber membrane separator with enhanced thermal stability and effective polysulfide inhibition for high-performance lithium–sulfur batteries. *J. Mater. Chem. A* **2017**, *5*, 15096–15104. [CrossRef]
83. Yanilmaz, M.; Lu, Y.; Dirican, M.; Fu, K.; Zhang, X. Nanoparticle-on-nanofiber hybrid membrane separators for lithium-ion batteries via combining electrospraying and electrospinning techniques. *J. Membr. Sci.* **2014**, *456*, 57–65. [CrossRef]
84. Zhang, F.; Ma, X.; Cao, C.; Li, J.; Zhu, Y. Poly(vinylidene fluoride)/SiO$_2$ composite membranes prepared by electrospinning and their excellent properties for nonwoven separators for lithium-ion batteries. *J. Power Sources* **2014**, *251*, 423–431. [CrossRef]
85. Zhang, S.; Cao, J.; Shang, Y.; Wang, L.; He, X.; Li, J.; Zhao, P.; Wang, Y. Nanocomposite polymer membrane derived from nano tio$_2$-pmma and glass fiber nonwoven: High thermal endurance and cycle stability in lithium ion battery applications. *J. Mater. Chem. A* **2015**, *3*, 17697–17703. [CrossRef]

86. Kim, Y.-S.; Xie, Y.; Wen, X.; Wang, S.; Kim, S.J.; Song, H.-K.; Wang, Z.L. Highly porous piezoelectric pvdf membrane as effective lithium ion transfer channels for enhanced self-charging power cell. *Nano Energy* **2015**, *14*, 77–86. [CrossRef]
87. Xing, L.; Nie, Y.; Xue, X.; Zhang, Y. Pvdf mesoporous nanostructures as the piezo-separator for a self-charging power cell. *Nano Energy* **2014**, *10*, 44–52. [CrossRef]
88. Gao, S.; Wang, K.; Wang, R.; Jiang, M.; Han, J.; Gu, T.; Cheng, S.; Jiang, K. Poly(vinylidene fluoride)-based hybrid gel polymer electrolytes for additive-free lithium sulfur batteries. *J. Mater. Chem. A* **2017**, *5*, 17889–17895. [CrossRef]
89. Shamshad, A.; Chao, T.; Muhammad, W.; Weiqiang, L.; Zhaohuan, W.; Songhao, W.; Bismark, B.; Jingna, L.; Junaid, A.; Jie, X.; et al. Highly efficient pvdf-hfp/colloidal alumina composite separator for high-temperature lithium-ion batteries. *Adv. Mater. Interfaces* **2018**, *5*, 1701147.
90. Yu, L.; Wang, D.; Zhao, Z.; Han, J.; Zhang, K.; Cui, X.; Xu, Z. Pore-forming technology development of polymer separators for power lithium-ion battery. In *Proceedings of Sae-China Congress 2014: Selected Papers*; Springer: Berlin, Germany, 2014; pp. 89–95.
91. Yeon, D.; Lee, Y.; Ryou, M.-H.; Lee, Y.M. New flame-retardant composite separators based on metal hydroxides for lithium-ion batteries. *Electrochim. Acta* **2015**, *157*, 282–289. [CrossRef]
92. Deng, Y.; Song, X.; Ma, Z.; Zhang, X.; Shu, D.; Nan, J. Al$_2$O$_3$/pvdf-hfp-cmc/pe separator prepared using aqueous slurry and post-hot-pressing method for polymer lithium-ion batteries with enhanced safety. *Electrochim. Acta* **2016**, *212*, 416–425. [CrossRef]
93. Liu, Y.; Qiao, Y.; Zhang, Y.; Yang, Z.; Gao, T.; Kirsch, D.; Liu, B.; Song, J.; Yang, B.; Hu, L. 3d printed separator for the thermal management of high-performance li metal anodes. *Energy Storage Mater.* **2018**, *12*, 197–203. [CrossRef]
94. Asghar, M.R.; Zhang, Y.; Wu, A.; Yan, X.; Shen, S.; Ke, C.; Zhang, J. Preparation of microporous cellulose/poly(vinylidene fluoride-hexafluoropropylene) membrane for lithium ion batteries by phase inversion method. *J. Power Sources* **2018**, *379*, 197–205. [CrossRef]
95. Kim, M.; Kim, J.K.; Park, J.H. Clay nanosheets in skeletons of controlled phase inversion separators for thermally stable li-ion batteries. *Adv. Funct. Mater.* **2015**, *25*, 3399–3404. [CrossRef]
96. Karuppasamy, K.; Reddy, P.A.; Srinivas, G.; Sharma, R.; Tewari, A.; Kumar, G.H.; Gupta, D. An efficient way to achieve high ionic conductivity and electrochemical stability of safer nonaflate anion-based ionic liquid gel polymer electrolytes (ilgpes) for rechargeable lithium ion batteries. *J. Solid State Electrochem.* **2017**, *21*, 1145–1155. [CrossRef]
97. Choi, Y.; Zhang, K.; Chung, K.Y.; Wang, D.H.; Park, J.H. Pvdf-hfp/exfoliated graphene oxide nanosheet hybrid separators for thermally stable li-ion batteries. *RSC Adv.* **2016**, *6*, 80706–80711. [CrossRef]
98. Xiao, W.; Wang, J.; Wang, H.; Gong, Y.; Zhao, L.; Liu, J.; Yan, C. Hollow mesoporous silica sphere-embedded composite separator for high-performance lithium-ion battery. *J. Solid State Electrochem.* **2016**, *20*, 2847–2855. [CrossRef]
99. Freitag, A.; Langklotz, U.; Rost, A.; Stamm, M.; Ionov, L. Ionically conductive polymer/ceramic separator for lithium-sulfur batteries. *Energy Storage Mater.* **2017**, *9*, 105–111. [CrossRef]
100. Suleman, M.; Kumar, Y.; Hashmi, S.A. High-rate supercapacitive performance of go/r-go electrodes interfaced with plastic-crystal-based flexible gel polymer electrolyte. *Electrochim. Acta* **2015**, *182*, 995–1007. [CrossRef]
101. Huang, F.; Liu, W.; Li, P.; Ning, J.; Wei, Q. Electrochemical properties of llto/fluoropolymer-shell cellulose-core fibrous membrane for separator of high performance lithium-ion battery. *Materials* **2016**, *9*, 75. [CrossRef] [PubMed]
102. Chen, W.; Liu, Y.; Ma, Y.; Liu, J.; Liu, X. Improved performance of pvdf-hfp/pi nanofiber membrane for lithium ion battery separator prepared by a bicomponent cross-electrospinning method. *Mater. Lett.* **2014**, *133*, 67–70. [CrossRef]
103. Kim, J.-H.; Kim, J.-H.; Choi, E.-S.; Kim, J.H.; Lee, S.-Y. Nanoporous polymer scaffold-embedded nonwoven composite separator membranes for high-rate lithium-ion batteries. *RSC Adv.* **2014**, *4*, 54312–54321. [CrossRef]
104. Padmaraj, O.; Rao, B.N.; Jena, P.; Venkateswarlu, M.; Satyanarayana, N. Electrospun nanocomposite fibrous polymer electrolyte for secondary lithium battery applications. In *AIP Conference Proceedings*; American Institute of Physics: College Park, MD, USA, 2014; Volume 1591, pp. 1723–1725.

105. Raja, M.; Angulakshmi, N.; Thomas, S.; Kumar, T.P.; Stephan, A.M. Thin, flexible and thermally stable ceramic membranes as separator for lithium-ion batteries. *J. Membr. Sci.* **2014**, *471*, 103–109. [CrossRef]
106. Zaccaria, M.; Fabiani, D.; Cannucciari, G.; Gualandi, C.; Focarete, M.L.; Arbizzani, C.; De Giorgio, F.; Mastragostino, M. Effect of silica and tin oxide nanoparticles on properties of nanofibrous electrospun separators. *J. Electrochem. Soc.* **2015**, *162*, A915–A920. [CrossRef]
107. Raja, M.; Kumar, T.P.; Sanjeev, G.; Zolin, L.; Gerbaldi, C.; Stephan, A.M. Montmorillonite-based ceramic membranes as novel lithium-ion battery separators. *Ionics* **2014**, *20*, 943–948. [CrossRef]
108. Xiao, W.; Gao, Z.; Wang, S.; Liu, J.; Yan, C. A novel naa-type zeolite-embedded composite separator for lithium-ion battery. *Mater. Lett.* **2015**, *145*, 177–179. [CrossRef]
109. Yang, C.-C.; Lian, Z.-Y. Electrochemical performance of $lini_{1/3}co_{1/3}mn_{1/3}o_2$ lithium polymer battery based on pvdf-hfp/m-sba15 composite polymer membranes. *Ceram. Trans.* **2014**, *246*, 18–202.
110. Yang, C.-C.; Lian, Z.-Y.; Lin, S.J.; Shih, J.-Y.; Chen, W.-H. Preparation and application of pvdf-hfp composite polymer electrolytes in $lini_{0.5}co_{0.2}mn_{0.3}o_2$ lithium-polymer batteries. *Electrochim. Acta* **2014**, *134*, 258–265. [CrossRef]
111. Kuo, P.L.; Tsao, C.H.; Hsu, C.H.; Chen, S.T.; Hsu, H.M. A new strategy for preparing oligomeric ionic liquid gel polymer electrolytes for high-performance and nonflammable lithium ion batteries. *J. Membr. Sci.* **2016**, *499*, 462–469. [CrossRef]
112. Wang, H.; Gao, H. A sandwich-like composite nonwoven separator for li-ion batteries. *Electrochim. Acta* **2016**, *215*, 525–534. [CrossRef]
113. Xiao, W.; Gong, Y.; Wang, H.; Liu, J.; Yan, C. Organic–inorganic binary nanoparticle-based composite separators for high performance lithium-ion batteries. *New J. Chem.* **2016**, *40*, 8778–8785. [CrossRef]
114. Yu, B.; Zhao, X.M.; Jiao, X.N.; Qi, D.Y. Composite nanofiber membrane for lithium-ion batteries prepared by electrostatic spun/spray deposition. *J. Electrochem. Energy Convers. Storage* **2016**, *13*, 1–6. [CrossRef]
115. Wang, M.; Chen, X.; Wang, H.; Wu, H.; Jin, X.; Huang, C. Improved performances of lithium-ion batteries with a separator based on inorganic fibers. *J. Mater. Chem. A* **2017**, *5*, 311–318. [CrossRef]
116. Suriyakumar, S.; Raja, M.; Angulakshmi, N.; Nahm, K.S.; Stephan, A.M. A flexible zirconium oxide based-ceramic membrane as a separator for lithium-ion batteries. *RSC Adv.* **2016**, *6*, 92020–92027. [CrossRef]
117. Solarajan, A.K.; Murugadoss, V.; Angaiah, S. Dimensional stability and electrochemical behaviour of zro_2 incorporated electrospun pvdf-hfp based nanocomposite polymer membrane electrolyte for li-ion capacitors. *Sci. Rep.* **2017**, *7*, 1–9. [CrossRef] [PubMed]
118. Li, X.; He, J.; Wu, D.; Zhang, M.; Meng, J.; Ni, P. Development of plasma-treated polypropylene nonwoven-based composites for high-performance lithium-ion battery separators. *Electrochim. Acta* **2015**, *167*, 396–403. [CrossRef]
119. Chen, W.; Liu, Y.; Ma, Y.; Yang, W. Improved performance of lithium ion battery separator enabled by co-electrospinnig polyimide/poly(vinylidene fluoride-co-hexafluoropropylene) and the incorporation of tio_2-(2-hydroxyethyl methacrylate). *J. Power Sources* **2015**, *273*, 1127–1135. [CrossRef]
120. Dai, M.; Shen, J.; Zhang, J.; Li, G. A novel separator material consisting of zeoliticimidazolate framework-4 (zif-4) and its electrochemical performance for lithium-ions battery. *J. Power Sources* **2017**, *369*, 27–34. [CrossRef]
121. Liu, J.; He, C.; He, J.; Cui, J.; Liu, H.; Wu, X. An enhanced poly(vinylidene fluoride) matrix separator with high density polyethylene for good performance lithium ion batteries. *J. Solid State Electrochem.* **2017**, *21*, 919–925. [CrossRef]
122. Li, H.; Niu, D.-H.; Zhou, H.; Chao, C.-Y.; Wu, L.-J.; Han, P.-L. Preparation and characterization of pvdf separators for lithium ion cells using hydroxyl-terminated polybutadiene grafted methoxyl polyethylene glycol (htpb-g-mpeg) as additive. *Appl. Surf. Sci.* **2018**, *440*, 186–192. [CrossRef]
123. Xiao, S.Y.; Yang, Y.Q.; Li, M.X.; Wang, F.X.; Chang, Z.; Wu, Y.P.; Liu, X. A composite membrane based on a biocompatible cellulose as a host of gel polymer electrolyte for lithium ion batteries. *J. Power Sources* **2014**, *270*, 53–58. [CrossRef]
124. Arbizzani, C.; Colò, F.; De Giorgio, F.; Guidotti, M.; Mastragostino, M.; Alloin, F.; Bolloli, M.; Molméret, Y.; Sanchez, J.Y. A non-conventional fluorinated separator in high-voltage graphite/$lini_{0.4}mn_{1.6}o_4$ cells. *J. Power Sources* **2014**, *246*, 299–304. [CrossRef]

125. Arbizzani, C.; De Giorgio, F.; Mastragostino, M. Characterization tests for plug-in hybrid electric vehicle application of graphite/lini$_{0.4}$mn$_{1.6}$o$_4$ cells with two different separators and electrolytes. *J. Power Sources* **2014**, *266*, 170–174. [CrossRef]
126. Wu, Q.Y.; Liang, H.Q.; Gu, L.; Yu, Y.; Huang, Y.Q.; Xu, Z.K. Pvdf/pan blend separators via thermally induced phase separation for lithium ion batteries. *Polymer* **2016**, *107*, 54–60. [CrossRef]
127. Zhu, Y.; Yin, M.; Liu, H.; Na, B.; Lv, R.; Wang, B.; Huang, Y. Modification and characterization of electrospun poly (vinylidene fluoride)/poly (acrylonitrile) blend separator membranes. *Composites Part B* **2017**, *112*, 31–37. [CrossRef]
128. Wu, X.; Liu, Y.; Yang, Q.; Wang, S.; Hu, G.; Xiong, C. Properties of gel polymer electrolytes based on poly(butyl acrylate) semi-interpenetrating polymeric networks toward li-ion batteries. *Ionics* **2017**, *23*, 2319–2325. [CrossRef]
129. Li, H.; Zhang, H.; Liang, Z.-Y.; Chen, Y.-M.; Zhu, B.-K.; Zhu, L.-P. Preparation and properties of poly (vinylidene fluoride)/poly(dimethylsiloxane) graft (poly(propylene oxide)-block-poly(ethylene oxide)) blend porous separators and corresponding electrolytes. *Electrochim. Acta* **2014**, *116*, 413–420. [CrossRef]
130. Kim, K.M.; Poliquit, B.Z.; Lee, Y.-G.; Won, J.; Ko, J.M.; Cho, W.I. Enhanced separator properties by thermal curing of poly(ethylene glycol)diacrylate-based gel polymer electrolytes for lithium-ion batteries. *Electrochim. Acta* **2014**, *120*, 159–166. [CrossRef]
131. La Monaca, A.; Arbizzani, C.; De Giorgio, F.; Focarete, M.L.; Fabiani, D.; Zaccaria, M. Electrospun membranes based on pvdf-peo blends for lithium batteries. *ECS Trans.* **2016**, *73*, 75–81. [CrossRef]
132. Monaca, A.L.; Giorgio, F.D.; Focarete, M.L.; Fabiani, D.; Zaccaria, M.; Arbizzani, C. Polyvinylidene difluoride–polyethyleneoxide blends for electrospun separators in li-ion batteries. *J. Electrochem. Soc.* **2017**, *164*, A6431–A6439. [CrossRef]
133. Zhu, C.; Nagaishi, T.; Shi, J.; Lee, H.; Wong, P.Y.; Sui, J.; Hyodo, K.; Kim, I.S. Enhanced wettability and thermal stability of a novel polyethylene terephthalate-based poly(vinylidene fluoride) nanofiber hybrid membrane for the separator of lithium-ion batteries. *ACS Appl. Mater. Interfaces* **2017**, *9*, 26400–26406. [CrossRef] [PubMed]
134. Park, S.; Son, C.W.; Lee, S.; Kim, D.Y.; Park, C.; Eom, K.S.; Fuller, T.F.; Joh, H.I.; Jo, S.M. Multicore-shell nanofiber architecture of polyimide/polyvinylidene fluoride blend for thermal and long-term stability of lithium ion battery separator. *Sci. Rep.* **2016**, *6*, 1–8. [CrossRef] [PubMed]
135. Yvonne, T.; Zhang, C.; Zhang, C.; Omollo, E.; Ncube, S. Properties of electrospun pvdf/pmma/ca membrane as lithium based battery separator. *Cellulose* **2014**, *21*, 2811–2818. [CrossRef]
136. Li, H.; Lin, C.-E.; Shi, J.-L.; Ma, X.-T.; Zhu, B.-K.; Zhu, L.-P. Preparation and characterization of safety pvdf/p(mma-co-pegma) active separators by studying the liquid electrolyte distribution in this kind of membrane. *Electrochim. Acta* **2014**, *115*, 317–325. [CrossRef]
137. Wu, X.L.; Lin, J.; Wang, J.Y.; Guo, H. Electrospun pvdf/pmma/sio$_2$ membrane separators for rechargeable lithium-ion batteries. *Key Eng. Mater.* **2015**, *645-646*, 1201–1206. [CrossRef]
138. He, T.; Jia, R.; Lang, X.; Wu, X.; Wang, Y. Preparation and electrochemical performance of pvdf ultrafine porous fiber separator-cum-electrolyte for supercapacitor. *J. Electrochem. Soc.* **2017**, *164*, 379–384. [CrossRef]
139. Gören, A.; Costa, C.M.; Tamaño Machiavello, M.N.; Cíntora-Juárez, D.; Nunes-Pereira, J.; Tirado, J.L.; Silva, M.M.; Gomez Ribelles, J.L.; Lanceros-Méndez, S. Effect of the degree of porosity on the performance of poly(vinylidene fluoride-trifluoroethylene)/poly(ethylene oxide) blend membranes for lithium-ion battery separators. *Solid State Ionics* **2015**, *280*, 1–9. [CrossRef]
140. Huang, F.; Xu, Y.; Peng, B.; Su, Y.; Jiang, F.; Hsieh, Y.-L.; Wei, Q. Coaxial electrospun cellulose-core fluoropolymer-shell fibrous membrane from recycled cigarette filter as separator for high performance lithium-ion battery. *ACS Sustain. Chem. Eng.* **2015**, *3*, 932–940. [CrossRef]
141. He, J.; Liu, J.; Li, J.; Lai, Y.; Wu, X. Enhanced ionic conductivity and electrochemical capacity of lithium ion battery based on pvdf-hfp/hdpe membrane. *Mater. Lett.* **2016**, *170*, 126–129. [CrossRef]
142. Farooqui, U.R.; Ahmad, A.L.; Hamid, N.A. Effect of polyaniline (pani) on poly(vinylidene fluoride-co-hexaflouro propylene) (pvdf-co-hfp) polymer electrolyte membrane prepared by breath figure method. *Polym. Test.* **2017**, *60*, 124–131. [CrossRef]
143. Kimura, N.; Sakumoto, T.; Mori, Y.; Wei, K.; Kim, B.-S.; Song, K.-H.; Kim, I.-S. Fabrication and characterization of reinforced electrospun poly(vinylidene fluoride-co-hexafluoropropylene) nanofiber membranes. *Compos. Sci. Technol.* **2014**, *92*, 120–125. [CrossRef]

144. Ding, G.; Qin, B.; Liu, Z.; Zhang, J.; Zhang, B.; Hu, P.; Zhang, C.; Xu, G.; Yao, J.; Cui, G. A polyborate coated cellulose composite separator for high performance lithium ion batteries. *J. Electrochem. Soc.* **2015**, *162*, A834–A838. [CrossRef]
145. Seidel, S.M.; Jeschke, S.; Vettikuzha, P.; Wiemhofer, H.D. Pvdf-hfp/ether-modified polysiloxane membranes obtained via airbrush spraying as active separators for application in lithium ion batteries. *Chem. Commun.* **2015**, *51*, 12048–12051. [CrossRef] [PubMed]
146. Freitag, A.; Stamm, M.; Ionov, L. Separator for lithium-sulfur battery based on polymer blend membrane. *J. Power Sources* **2017**, *363*, 384–391. [CrossRef]
147. Zhai, Y.; Xiao, K.; Yu, J.; Ding, B. Closely packed x-poly(ethylene glycol diacrylate) coated polyetherimide/poly(vinylidene fluoride) fiber separators for lithium ion batteries with enhanced thermostability and improved electrolyte wettability. *J. Power Sources* **2016**, *325*, 292–300. [CrossRef]

© 2018 by the authors. Licensee MDPI, Basel, Switzerland. This article is an open access article distributed under the terms and conditions of the Creative Commons Attribution (CC BY) license (http://creativecommons.org/licenses/by/4.0/).

Article

Gel Polymer Electrolytes Based on Silica-Added Poly(ethylene oxide) Electrospun Membranes for Lithium Batteries

Maria Assunta Navarra [1,*], Lucia Lombardo [1], Pantaleone Bruni [2], Leonardo Morelli [1], Akiko Tsurumaki [1], Stefania Panero [1] and Fausto Croce [2,*]

1. Dipartimento di Chimica, Sapienza Università di Roma, Piazzale Aldo Moro 5, 00185 Rome, Italy; lucia.lombardo@uniroma1.it (L.L.); morelli.1424793@studenti.uniroma1.it (L.M.); akiko.tsurumaki@uniroma1.it (A.T.); stefania.panero@uniroma1.it (S.P.)
2. Dipartimento di Farmacia, Università "G. d'Annunzio" Chieti-Pescara, Via dei Vestini 31, 66100 Chieti, Italy; pantaleone.bruni@unich.it
* Correspondence: mariassunta.navarra@uniroma1.it (M.A.N.); fausto.croce@unich.it (F.C.); Tel.: +39-06-4991-3658 (M.A.N.); +39-0871-355-4480 (F.C.)

Received: 17 October 2018; Accepted: 30 November 2018; Published: 5 December 2018

Abstract: Solid polymer electrolytes, in the form of membranes, offering high chemical and mechanical stability, while maintaining good ionic conductivity, are envisaged as a possible solution to improve performances and safety in different lithium cell configurations. In this work, we designed and prepared systems formed using innovative nanocomposite polymer membranes, based on high molecular weight poly(ethylene oxide) (PEO) and silica nanopowders, produced by the electrospinning technique. These membranes were subsequently gelled with solutions based on aprotic ionic liquid, carbonate solvents, and lithium salt. The addition of polysulfide species to the electrolyte solution was also considered, in view of potential applications in lithium-sulfur cells. The morphology of the electrospun pristine membranes was evaluated using scanning electron microscopy. Stability and thermal properties of pristine and gelled systems were investigated uisng differential scanning calorimetry and thermal gravimetric analysis. Electrochemical impedance spectroscopy was used to determine the conductivity of both swelling solutions and gelled membranes, allowing insight into the ion transport mechanism within the proposed composite electrolytes.

Keywords: gel polymer electrolyte; electrospinning; lithium batteries

1. Introduction

Intense research efforts are still directed to improve the characteristics of lithium battery devices for energy storage. In particular, attention is directed to those devices that possess the requirements of safety and performance for powering electric vehicles [1] or for stationary use in smart-grids [2,3]. In this scenario, great advances in terms of reliability can be achieved by moving from liquid to polymer electrolytes. Among the different types of polymer electrolytes for lithium batteries, those based on polyethylene oxide (PEO) membranes, have received significant attention [4].

PEO membranes combine high chemical stability, solid-like diffusivity, and good ionic conductivity [5,6]. The main problem that has limited the use of PEO is its poor ionic conductivity at room temperature (10^{-8} S/cm) due to the presence of crystalline phases. Conductivity values of interest for applications in lithium batteries (>10^{-4} S/cm) are reached only above 65 °C, i.e., beyond the melting temperature of the polymer crystalline phase. The reason for this behavior is to be found in the peculiar PEO conduction mechanism as the lithium ion moves from a coordination center (the PEO oxygen atoms) to the next. This movement is made possible only if the chain undergoes subsequent rearrangement steps. The dynamics of this process is greatly facilitated by a high mobility of the chains. Consequently, high values of ionic conductivity are reached above the melting temperature in the amorphous phase where long-range chain active motions are possible [7].

In this paper we address the problem of PEO low room temperature conductivity using three different strategies [8]. The first one consists of incorporating inside the fibers of electrospun PEO membranes inert ceramic particles, which act as fillers. The presence of these particles inhibits the PEO chains tendency to crystallize, leading to membranes with a higher amorphous fraction. In addition, as is well documented in the literature, it is known that the fillers stabilize the lithium electrolyte interface and increase the Li$^+$ transference number due to Lewis acid–base type interactions between the ceramic surface groups and the polymer chains coordination sites. As an additional benefit, the presence of this inorganic dispersion improves the mechanical properties of the composite polymer membranes. In general, the best overall performances have been obtained with nanometric particles having acidic surface groups and in a 5–20 wt.% ratio with respect to the polymer [5,8–11]. The second strategy involves the use of lithium salts with low lattice energy, which favors the lithium salt dissociation. To achieve this goal, lithium salts with large and flexible anions, that disperse effectively the charge and are able to increase the free-volume between the polymer chains and thus to promote the ionic mobility, have been used [8,12,13]. The third approach has involved the addition of an aprotic ionic liquid (IL) that increases the total ionic conductivity, competing with lithium-ions toward the binding sites of coordination and creating additional free-volume between the chains [8]. The disadvantages of ionic liquids are their high viscosity, their tendency to form ionic clusters, and their high cost [14]. To mitigate such drawbacks, we have added to the electrolyte formulations alkyl carbonates, which reduce the viscosity of the solution and form a protective passivating film (i.e., a solid–electrolyte interface, SEI) on the lithium anode [8].

In this perspective, we designed and prepared innovative electrolytic systems with the aim of providing a compositional study and elucidation on the role of different components. Emphasis is given to the analysis of the conducting properties of the proposed systems as a figure of merit in view of applications in electrochemical devices. Such electrolytes consist of nanocomposite polymer membranes of PEO/silica produced through the electrospinning technique and then gelled using two liquid solutions. The two solutions, based on the aprotic ionic liquid PYR$_{14}$TFSI, contain a mixture of two carbonate solvents, i.e., ethylene carbonate and dimethyl carbonate (EC and DMC), and the bis(trifluoromethane)sulfonyl imide lithium salt (LiTFSI). In one of the two solutions, the polysulfide Li$_2$S$_8$ was also added. The addition of this last component is justified by the fact that these systems have been thought to provide possible future applications not only in conventional lithium batteries but, specifically, in lithium-sulfur batteries [15–17]. The dissolution of the sulfur cathode in the form of polysulfides (and the consequent shuttle phenomenon) strongly limits the development of this technology. For this reason, we intend to reduce such dissolution playing on the solubility equilibrium, by adding a buffer polysulfide to the electrolyte [18–20]. Also, some ILs (such as N-methyl-N-butylpyrrolidinium bis(trifluoromethanesulfonyl)imide, PYR$_{14}$TFSI) have already demonstrated the ability to suppress the dissolution of polysulfides increasing the performance of sulfur cathodes [16]. The choice of LiTFSI as lithium source is due to its stability, large dimension, and flexible imide structure. These latter properties, together with the strong electron withdrawing behavior of the (trifluoromethane)sulfonyl group, enhance the negative charge delocalization and, in turn, guarantee a high salt dissociation level. Moreover, the TFSI anion is the same contained in the ionic liquid, which should avoid the Li$^+$-ion transference number reduction, possibly occurring when additional ionic species are added to the solution [21].

2. Experimental

2.1. Materials for Membranes Preparation

Poly(ethylene oxide) (PEO, Mw = 4 M ≈ 4×10^6 g/mol) and Silica (Nanopowder, average size = 10–20 nm—"BET" Brunauer–Emmett–Teller) were purchased form Sigma-Aldrich (St. Louis, MO, USA). Bi-distilled water was produced in house.

2.2. Membrane Separators Fabrication

An in-house made electrospinning apparatus, composed of a high voltage power supply (Spellman SL 50 P 10/CE/230, Hauppauge, NY, USA), a syringe pump (KD Scientific 200 series, Holliston, MA, USA), a glass syringe, a stainless steel blunt-ended needle (inner diameter: 0.84 mm) connected with the power supply electrode, and a grounded aluminum plate-type collector (area ≈10 cm^2), was utilized in order to prepare the electrospun PEO membranes. The water-based PEO polymer solution was dispensed through a Teflon tube to the needle that was vertically placed on the collecting plate. PEO was dissolved at a concentration of 3% w/v in bi-distilled water. To create the composite systems, after the polymer dissolution, silica nanoparticles were added to the resulting solutions in proper amounts in order to produce a final membrane containing 10% w/w of silica. The realized dispersions were electrospun by using the following conditions: voltage = 13 kV, needle-to-collector distance = 30 cm, and flow rate = 0.005 mL/min. A quality control was performed in order to verify the homogeneity of the sample thickness by means of a digital micrometer. Only specimens having thicknesses in the range of 50 ± 10 μm were used for further characterizations.

The two synthesized membranes will be here labelled as "es-PEO" (for the electrospun sample based on pure PEO) and "es-PEO-SiO$_2$" (for the electrospun sample based on PEO with 10% of silica particles)

2.3. Swelling Solutions

Liquid electrolyte solutions were prepared as gelling media of the polymer membranes, in order to evaluate the potentialities of the resulting gelled membranes as polymer electrolytes for lithium batteries. In particular, in view of their potential application in lithium-sulfur batteries, two similar solutions, with and without a polysulfide (henceforward called "PS-containing sol" and "PS-free sol," respectively), were prepared. PS-containing sol was formed using PyR$_{14}$TFSI 77 wt.%, EC:DMC (1:1 volume ratio) 18.4 wt.%, LiTFSI 0.5 mol/kg, Li$_2$S$_8$ 4.6 wt.%, and PS-free sol used PyR$_{14}$TFSI 80 wt.%, EC:DMC (1:1 volume ratio) 20 wt.%, and LiTFSI 0.5 mol/kg.

The ionic liquid N-methyl-N-butylpyrrolidinium bis(trifluoromethanesulfonyl)imide (PYR$_{14}$TFSI) was bought from Solvionic (99.9%, Toulouse, France) and used as received; the lithium salt (bis(trifluoromethane)sulfonyl imide (LiTFSI) was purchased from Fluka (99.9%, Munich, Germany) and used without further purification; carbonates solvents, i.e., ethylene carbonate (EC) and dimethyl carbonate (DMC), were provided by Sigma-Aldrich. Polysulfide (in form of Li$_2$S$_8$) was produced in situ via direct reactions between metallic lithium (Chemetall, Frankfurt am Main, Germany) and elemental sulfur (Sigma-Aldrich) in stoichiometric ratio, in a corked vial where the other components were already presents; the mixture was heated to 60–70 °C, stirred for the first four hours, and let it stand for the next twenty hours. These operations were performed in a controlled argon atmosphere dry box having a humidity content below 1 ppm.

2.4. Characterizations

The PEO-based membranes were characterized, both dry and after gelation by the two liquid electrolytes.

Gelation of membranes was realized by dropping the liquid electrolyte on disks of membranes, keeping the weight ratio solution-to-membrane ≈5.

The morphology of the electrospun pristine dry membranes was evaluated by means of scanning electron microscopy (SEM—EVO50, Zeiss, Jena, Germany). In order to reduce the charge accumulation, the membranes were covered with a thin layer of evaporated gold before SEM measurements.

Thermal properties were evaluated by means of differential scanning calorimetry (DSC) and thermal gravimetric analysis (TGA). DSC measurements on both pristine and gelled membranes, as well as on pure PEO powder, were performed with a Mettler-Toledo DSC 821 (Zaventem, Belgium) instrument under an inert nitrogen flux, cooling from room temperature down to $-90\,°C$, holding for 10 min at $-90\,°C$, and then heating up to $80\,°C$ at a scan rate of $10\,°C/min$. TGA was carried out on pristine membrane samples and on pure PEO powder, with a Mettler-Toledo TGA/SDTA 851e under an inert nitrogen flux, heating from room temperature to $600\,°C$ at a scan rate of $10\,°C/min$.

Electrochemical impedance spectroscopy (EIS) was used to determine the conductivity of the two swelling solutions and of the gelled membranes. To obtain the conductivity values of the solutions, the measurements were carried out by dipping a conductivity cell for liquids (composed of two sheets of platinum facing at the distance of 1 cm) in the test solution and controlling the temperature with a oven (Büchi-Oven B-585, BUCHI Italia S.r.l., Cornaredo, Italy) in the range 20–60 °C. A VSP potentiostat/galvanostat (Bio-Logic Science Instruments, Seyssinet-Pariset, France) was used to record the impedance spectra of the samples in the frequency range of 1 mHz–1 kHz with a sinusoidal signal of 5 mV amplitude. All the spectra, plotted as Nyquist plots, showed an almost vertical straight line intercepting the real axis at high frequency. The value in Ω of this intercept, i.e., the cell resistance at infinite frequency, has been used to calculate the conductivity of the samples under study through the equation:

$$\sigma = 1/R \times (L/S),$$

where "L" and "S" are, respectively, the distance in centimeters and the surface area in centimeters square of the electrodes.

The conductivity of gelled membranes was evaluated using EIS, assembling coin-type cells with stain-less steel current collector electrodes, where the swollen PEO membrane acts as an electrolyte separator. For each cell, a 100 µm-thick Teflon O-ring spacer was adopted and two disks of membrane having diameter 0.8 cm were directly gelled in the cell by dropping the desired amount of liquid electrolyte. Impedance spectra were recorded by applying a 10 mV amplitude signal in the frequency range 200 kHz–1 kHz using a VMP2 potentiostat/galvanostat (Bio-Logic Science Instruments, Seyssinet-Pariset, France). The temperature was controlled in the range $-50\,°C$ to $+60\,°C$ with a Tenney Junior Compact Temperature Test Chamber (TPS, White-Deer, PA, USA).

3. Results and Discussion

Morphology of pristine PEO-based electrospun membranes is shown in Figure 1. SEM images were recorded for both es-PEO (Figure 1a) and es-PEO-SiO$_2$ (Figure 1b) samples, without and with the inorganic additive, respectively. All samples are made of bead-free fibers with an average diameter of 250–300 nm. When nanoparticles were added to PEO solution, the resulting fibers showed slightly higher diameter that can be explained by the inclusion of silica particles inside the polymer fibers. Moreover, Figure 1b shows a certain degree of silica agglomerates, deposited on single fibers or within voids in the polymer mat. Such micrometric agglomerates resulted from the aggregation during the electrospinning process in aqueous media of silica particles, given as spherical porous nanopowder by the product specification.

Thermal stability of pristine membranes, compared to pure PEO powder, was investigated by TGA measurements, as shown in Figure 2.

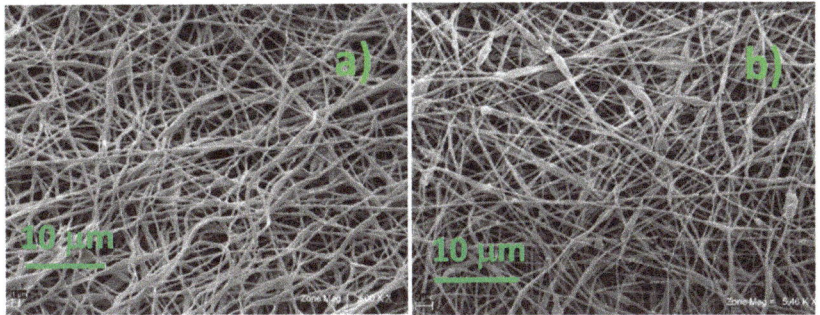

Figure 1. SEM images of the two electrospun membranes: (**a**) without silica additive, and (**b**) with 10 wt.% SiO$_2$.

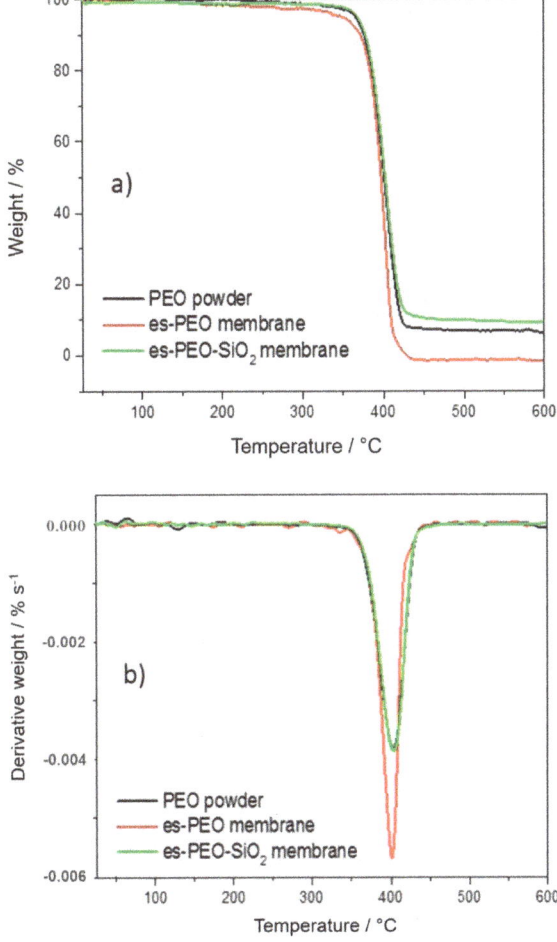

Figure 2. TGA (**a**) and derivative-TGA (**b**) curves of the two electrospun pristine membranes and of pure PEO powder.

45

High thermal stability typical of pure PEO polymer, extending up to 400 °C, is preserved in both pristine membranes. As highlighted by the minimum of the derivative curves in Figure 2b, temperatures of decomposition were quite similar for all the investigated samples (i.e., 401 °C for the es-PEO membranes and ≈404 °C for both PEO powder and es-PEO-SiO$_2$). Unexpected residual masses were revealed above 450 °C in Figure 2a. Pure PEO powder was not completed removed beyond its decomposition temperature, giving ≈8% of left over weight. This could be attributed to non-volatile residuals and to the presence of thermo-stable additives or catalytic compounds used in the synthesis of the polymer. It should be noticed that such residual mass is expected to be reduced by lowering the heating rate, due to a diffusion-limited elimination of the decomposition products by the powder bulk. This did not happen in the case of es-PEO membrane, showing no residuals after it decomposes. Products of decomposition were easily removed from this high surface area sample and eventual additives, such as inhibitors and stabilizers, present in the starting powder were separated and eliminated during the electro-spinning process in aqueous media. A residual mass slightly higher than 10% was revealed for the es-PEO-SiO$_2$ membrane, which was attributed to the inorganic silica filler.

The DSC response of the two pristine membranes, compared with pure PEO powder, is displayed in Figure 3. One main thermal transition was evident in all the DSC traces due to the melting of the polymer crystalline phase. The temperature and energy involved in this melting process have been evaluated and reported in Table 1. Temperature values here shown have been derived from the minimum of the endothermic peak, and in this respect, no valuable differences are noticed among the samples. The additive-free electrospun membrane is actually the sample showing a slightly lower melting temperature and a narrower peak. More remarkable differences were observed in terms of the enthalpy change, highlighting the role of both the electrospinning process and the silica filler. Clearly, lower energy was involved in the melting of pure PEO powder, which also revealed its lower crystallinity. The electrospinning process, due to its ordering effect, somehow increased the crystalline degree of the membranes, compared to the starting polymer powder, corresponding to a higher melting enthalpy. On the contrary, silica nanoparticles and their micrometric aggregates have the effect of lowering crystallinity, towards a more amorphous system with respect to plain, PEO-based electrospun membranes.

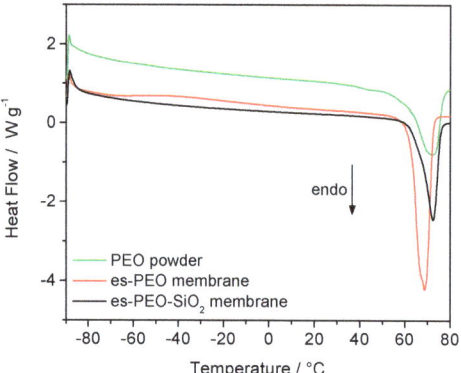

Figure 3. Heating scan of the DSC curves recorded for the two electrospun pristine membranes and for pure PEO powder.

Table 1. Enthalpy change and temperature related to PEO melting (values derived from DSC response in Figure 3).

Sample	$\Delta H_{melting}/J\cdot g^{-1}$	$T_{melting}/°C$
PEO powder	104.0	72.3
es-PEO membrane	188.5	68.7
es-PEO-SiO$_2$ membrane	149.7	72.2

Gelation of the polymer membranes using liquid electrolyte solutions (i.e., PS-free sol and PS-containing sol) was achieved to finally obtain the desired Li$^+$-conducting composite electrolytes. The transition from solid to gel-like systems was very easily and quickly attained with the proposed electrospun fiber mats due to their high surface-to-volume ratio. Thermal properties of the resulting electrolytes have been checked after the gelation process. The DSC response of the new gelled electrolytes is reported in Figure 4. As expected, the melting transition of the polymer was highly influenced by the presence of the liquid component. The crystallinity of PEO was strongly reduced, almost suppressed, upon gelation, giving rise to an amorphous, plasticized electrolyte system. In this respect, the nature of the electrolyte solution, with or without polysulfide, appeared almost irrelevant. The main role was due to the ionic liquid (i.e., the major component of the liquid electrolytes), which interacted with the polymer chains, thus preventing their crystallization. Interestingly, a certain degree of crystalline phase, even though very small, was preserved when silica particles were present (see the melting peak around 60 °C in Figure 4b). This is quite reasonably attributed to preferential interactions established between the inorganic filler and the liquid solution, leaving partially-coordinated PEO chains free to crystallize. In this respect, if we assume the starting PEO powder as a reference, it was possible to compare the crystallinity of our different samples by dividing the enthalpy change for the melting transition of each quoted sample by the enthalpy of melting related to the PEO powder. As already pointed out, crystallinity of the electrospun starting membranes was higher than that of the PEO powder due to the ordering effect of the electrospinning process [3], i.e., 1.81 times higher for the es-PEO membrane and 1.44 times higher for the es-PEO-SiO$_2$ membrane. Such an estimate was not possible for silica-free gelled systems, as no melting was observed in the DSC traces. Whereas, a very low crystalline degree was maintained in the silica-added gel polymer electrolytes, with the crystallinity being 0.12 and 0.08 times that of PEO powder for es-PEO-SiO$_2$/PS-free sol and es-PEO-SiO$_2$/PS-containing sol, respectively. Another thermal response was noticed around −80 °C in the DSC traces of Figure 4 due to the glass transition of the ionic liquid component [22].

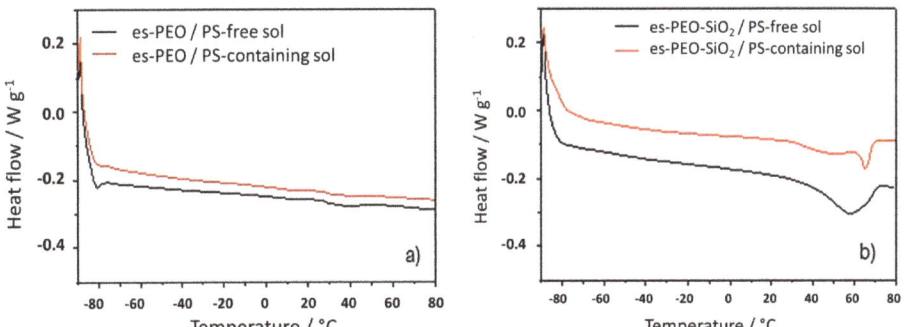

Figure 4. Heating scan of the DSC response of additive free PEO membrane (**a**) and of silica-added PEO membrane (**b**) gelled using liquid electrolyte "PS-free sol" (black curves) or "PS-containing sol" (red curves).

Functionality of the gelled membranes as electrolyte was tested using EIS measurements performed at increasing temperature in the range −50 °C to 60 °C. For comparison purposes, impedance spectra were recorded for the liquid electrolyte solutions as well, in this case limiting the temperature range between 20 °C and 70 °C. Conductivity values, obtained from the impedance spectra in the investigated T-ranges, are reported in Figure 5a,b for the swelling solutions and for the swollen membranes, respectively, in the form of Arrhenius plots. Conductivity values, extrapolated from the plots in Figure 5b at two temperatures of interest (i.e., 25 °C and 50 °C, representing normal operating conditions and melting region of the polymer, respectively) are reported in Table 2.

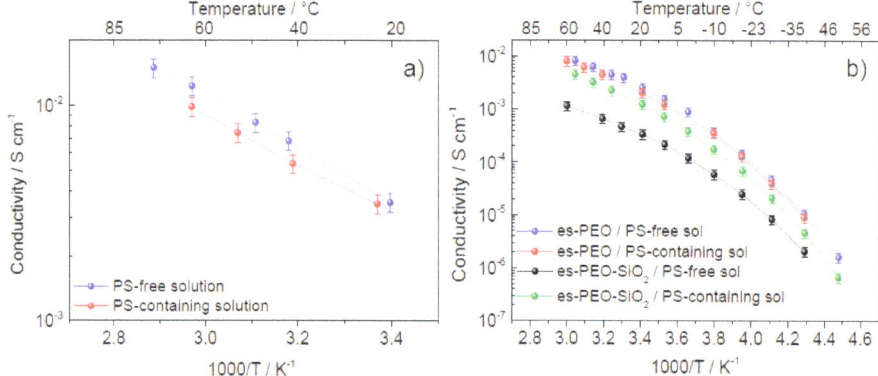

Figure 5. Arrhenius plots of conductivity of the swelling liquid solution (**a**) and of the gelled electrolyte membranes (**b**).

Table 2. Conductivity values extrapolated at 25 °C and 50 °C for the gelled electrolyte membranes.

Sample	$\sigma_{T = 25\,°C}$ (S/cm)	$\sigma_{T = 50\,°C}$ (S/cm)
es-PEO/PS-free sol	3.3×10^{-3}	7.2×10^{-3}
es-PEO/PS-containing sol	2.5×10^{-3}	6.1×10^{-3}
es-PEO-SiO$_2$/PS-free sol	4.0×10^{-4}	8.8×10^{-4}
es-PEO-SiO$_2$/PS-containing sol	1.5×10^{-3}	3.7×10^{-3}

With the exception of the es-PEO-SiO$_2$/PS-free sol sample, very high conductivities were achieved for all the gelled systems at both investigated temperatures. It is worth noticing that these room-temperature σ values are typical of viscous organic electrolytes used in lithium-batteries, revealing that the polymer membranes here were very well plasticized. As shown below, a very limited conductivity decrease was observed when moving from the liquid electrolytes, PS-free sol and PS-containing sol, to the gelled polymer systems. As expected, a temperature-activated transport mechanism was found, giving higher conductivity values at 50 °C with respect to 25 °C.

In Figure 5a, it is possible to observe both the liquid electrolytes showing interesting ionic conductivity, and the presence of polysulfide ions (in PS-containing solution) seemed to affect the conduction properties very little. Overall, the detected conductivity values appear suitable for battery applications in a wide temperature range. The addition of polysulfide in solution has, in general, two opposite effects on the overall ionic conductivity: on one hand, the number of charged species increased due to the intake of anions in the solution; on the other hand, an increase in viscosity is expected, which hindered the ion mobility. In our systems, the two solutions had very high, similar conductivity values, indicating that these effects offset each other.

As shown in Figure 5b, conductivity was lower for the silica-added gel polymer electrolytes (black and green plots) compared to the silica-free systems (blue and red plots). This can be explained by considering a possible retention of the liquid electrolyte fraction on the silica particles, hindering the polymer chain-assisted ion transport. In this respect, a certain role is played by the nature of the liquid electrolyte. Indeed, big differences were observed among the two silica-added membranes according to the type of swelling liquid electrolyte, with the polysulfide-doped solution giving higher overall conductivity with respect to the polysulfide-free system (compare es-PEO-SiO$_2$/PS-containing sol with es-PEO-SiO$_2$/PS-free sol in Figure 5b).

Each curve of Figure 5b was fitted using the Vogel–Tamman–Fulcher (VTF) equation:

$$\sigma(T) = A \cdot T^{-\frac{1}{2}} \cdot e^{-\frac{E_A}{R(T-T_0)}}$$

with A and E_A parameters representing the charge carrier number and the activation energy, respectively, whereas T_0 is the ideal glass transition temperature. Very high correlation (higher than 0.99) was found between the experimental and fitted curves, revealing that our systems followed the typical behavior of ion-conducting amorphous matrices where the polymer component assisted the ion transport. The extrapolated parameters were considered and reported in Table 3.

Table 3. Parameters of the VTF equation derived by fitting the curves in Figure 5b.

Sample	E_A (kJ/mol)	T_0 (K)	A (S/cm)
es-PEO/PS-free sol	6.76 ± 0.4	164.3 ± 2.5	23.27 ± 5.67
es-PEO/PS-containing sol	5.92 ± 0.3	170.5 ± 2.3	11.84 ± 2.38
es-PEO-SiO$_2$/PS-free sol	5.05 ± 0.2	170.7 ± 1.9	1.24 ± 0.18
es-PEO-SiO$_2$/PS-containing sol	7.00 ± 0.1	163.6 ± 0.8	13.80 ± 1.06

E_A values were low, compared to other IL-added PEO-based ion-conducting systems [23], meaning that ion transport and conduction mechanism were very easily activated. Moreover, no substantial differences were found among samples in terms of activation energy. Similarly, comparably low values of T_0 were obtained, proving that all the gelled electrolytes exhibited amorphous behavior. Differences of relevance were found in terms of the A parameter. The smallest value was obtained for es-PEO-SiO$_2$/PS-free sol sample, revealing that its low conductivity (see black curve in Figure 5b and values in Table 2) was actually due to a reduced number of total charge carriers. This supports our hypothesis that silica particles absorbed the liquid electrolyte, thus limiting the concentration of ions available for transport. Based on the higher conductivity and A values observed in es-PEO-SiO$_2$/PS-containing sol, we conclude that the polysulfide opposes this retention effect of SiO$_2$ additive to the advantage of the transport mechanism.

4. Conclusions

An easy way to obtain highly conductive gel polymer electrolytes was proposed in this paper. The swelling ability of high surface area, electrospun PEO membranes was exploited for the absorption of stable IL-based electrolytes. The gelling procedure, in terms of solid-to-liquid weight ratio (i.e., PEO membrane: IL-based electrolyte), was optimized to obtain a reproducible and high swelling degree. The absorption of the liquid component into the PEO membrane strongly reduced or even suppressed the crystallinity of the polymer, giving rise to amorphous, well-plasticized electrolyte systems. The effect of a silica particle additive, dispersed in the polymer matrices during the electrospinning process, was also investigated. A certain degree of crystalline phase, even though very small, was preserved in the membranes after gelling when silica was present, which was attributed to preferential interactions between the inorganic filler and the liquid solution, leaving partially-coordinated PEO chains free to crystallize. This was reflected in the conducting properties of the gel polymer electrolytes, showing lower ion conductivity in the silica-added sample because of a reduced number of available charge carriers.

In view of possible applications in lithium-sulfur batteries, the addition of a polysulfide component (i.e., Li$_2$S$_8$) in the swelling solution was considered. Such a polysulfide affected the ionic conductivity of the silica-free gel polymer electrolytes very little. On the contrary, it had a beneficial effect in the presence of silica. Apparently, the polysulfide opposed the liquid retention of SiO_2 particles to the advantage of the transport mechanism.

Overall, selected compositions proposed here showed conductivity values suitable for battery applications in a wide range of temperatures. All these findings address the potentiality of such gelled electrolytes.

Author Contributions: M.A.N. and F.C. conceived the work and designed the experiments; L.M. and P.B. synthesized materials and performed the experiments; L.L. and A.T. analyzed the data; M.A.N., S.P. and F.C. revised data analysis and discussion of results; M.A.N. and F.C. wrote the paper.

Funding: The results of this work have been obtained with the financial support of the European Community within the Seventh Framework Program LISSEN (Lithium Sulfur Superbattery Exploiting Nanotechnology) Project (project number 314282).

Conflicts of Interest: The authors declare no conflict of interest.

References

1. Opitz, A.; Badami, P.; Shen, L.; Vignarooban, K.; Kannan, A.M. Can Li-Ion batteries be the panacea for automotive applications? *Renew. Sust. Energ. Rev.* **2017**, *68*, 685–692. [CrossRef]
2. Ahmadi, L.; Young, S.B.; Fowler, M.; Fraser, R.A.; Achachlouei, M.A. A cascaded life cycle: Reuse of electric vehicle lithium-ion battery packs in energy storage systems. *Int. J. Life Cycle Assess.* **2017**, *22*, 111–124. [CrossRef]
3. Focarete, M.L.; Croce, F.; Scrosati, B.; Hassoun, J.; Meschini, I. A safe, high-rate and high energy polymer lithium-ion battery based on gelled membranes prepared by electrospinning. *Energy Envoron. Sci.* **2011**, *4*, 921–927.
4. Long, L.; Wang, S.; Xiao, M.; Meng, Y. Polymer electrolytes for lithium polymer batteries. *J. Mater. Chem. A* **2016**, *4*, 10038–10069. [CrossRef]
5. Croce, F.; Scrosati, B. Nanocomposite lithium ion conducting membranes. *Ann. N Y. Acad. Sci.* **2003**, *984*, 194–207. [CrossRef] [PubMed]
6. Agrawal, R.C.; Pandey, G.P. Solid Polymer electrolyte: Materials designing and all-solid-state battery applications. An overview. *J. Phys. D Appl. Phys.* **2008**, *41*, 223001. [CrossRef]
7. Devaux, D.; Bouchet, R.; Glé, D.; Denoyel, R. Mechanism of ion transport in PEO\LiTFSI complexes: Effect of temperature, molecoular weight and end groups. *Solid State Ionics* **2012**, *227*, 119–127. [CrossRef]
8. Wetjen, M.; Navarra, M.A.; Panero, S.; Passerini, S.; Scrosati, B.; Hassoun, J. Composite Poly(ethylene oxide) Electrolytes Plasticized by N-Alkyl-N-butylpyrrolidinium Bis(trifluoromethane-sulfonyl)imide for Lithium Batteries. *ChemSusChem* **2013**, *6*, 1037–1043. [CrossRef] [PubMed]
9. Croce, F.; Curini, R.; Martinelli, A.; Persi, L.; Ronci, F.; Scrosati, B.; Caminiti, R. Physical and Chemcal Properties of Nanocomposite Polymer Electrolytes. *J. Phys. Chem. B* **1999**, *103*, 10632–10638. [CrossRef]
10. Croce, F.; Appetecchi, G.B.; Persi, L.; Scrosati, B. Nanocomposite Polymer electrolytes for lithium batteries. *Nature* **1998**, *394*, 456–458. [CrossRef]
11. Zaccaria, M.; Gualandi, C.; Fabiani, D.; Focarete, M.L.; Croce, F. Effect of Oxide Nanoparticles on Thermal and Mechanical Properties of Electrospun Separators for Lithium-Ion Batteries. *J. Nanomater.* **2012**, *2012*, 119. [CrossRef]
12. Scrosati, B.; Garche, J. Lithium batteries: Status, prospects and future. *J. Power Sources* **2009**, *195*, 2419–2430. [CrossRef]
13. Shin, J.H.; Henderson, W.A.; Tizzani, C.; Passerini, S.; Jeong, S.S.; Kim, K.W. Characterization of Solvent-Free Polymer Electrolytes Consisting in Ternary PEO-LiTFSI-PYR14TFSI. *J. Electrochem. Soc.* **2006**, *153*, 1649–1654. [CrossRef]
14. Navarra, M.A. Ionic liquids as safe electrolyte components for Li-metal and Li-Ion batteries. *Mater. Res. Bull.* **2013**, *38*, 548–553. [CrossRef]
15. Chen, R.; Zhao, T.; Wu, F. From a historic review to horizons beyond: Lithium–sulphur batteries run on the wheels. *Chem. Commun.* **2015**, *51*, 18–33. [CrossRef] [PubMed]
16. Chen, L.; Shaw, L.L. Recent Advances in Lithium-Sulfur Batteries. *J. Power Sources* **2014**, *267*, 770–783. [CrossRef]

17. Lin, Z.; Liang, C. Lithium-sulfur batteries: From liquid to solid cells. *J. Mater. Chem. A* **2015**, *3*, 936–958. [CrossRef]
18. Chen, S.; Dai, F.; Gordin, M.L.; Wang, D. Exceptional electrochemical performance of rechargeable Li-S batteries with a polysulfide-containing electrolyte. *RSC Adv.* **2013**, *3*, 3540–3543. [CrossRef]
19. Ji, X.; Evers, S.; Black, R.; Nazar, L.F. Stabilizing lithium-sulphur cathodes using polysulphide reservoirs. *Nat. Commun.* **2011**, *2*, 325. [CrossRef]
20. Xu, R.; Belharouak, I.; Li, J.C.; Zhang, X.; Bloom, I.; Bareño, J. Role of Polysulfides in self-Healing Lithium-Sulfur Batteries. *Adv. Energy Mater.* **2013**, *3*, 833–838. [CrossRef]
21. Younesi, R.; Veith, G.M.; Johansson, P.; Edstrom, K.; Vegge, T. Lithium salts for advanced lithium batteries: Li–metal, Li–O2, and Li–S. *Energy Environ. Sci.* **2015**, *8*, 1905–1922. [CrossRef]
22. Martinelli, A.; Matic, A.; Jacobsson, P.; Börjesson, L. Phase Behavior and Ionic Conductivity in Lithium Bis(trifluoromethanesulfonyl)imide-Doped Ionic Liquids of the Pyrrolidinium Cation and Bis(trifluoromethanesulfonyl)imide Anion. *J. Phys. Chem. B* **2009**, *113*, 11247–11251. [CrossRef] [PubMed]
23. Cheng, H.; Zhu, C.; Huang, B.; Lu, M.; Yang, Y. Synthesis and electrochemical characterization of PEO-based polymer electrolytes with room temperature ionic liquids. *Electrochim. Acta* **2007**, *52*, 5789–5794. [CrossRef]

 © 2018 by the authors. Licensee MDPI, Basel, Switzerland. This article is an open access article distributed under the terms and conditions of the Creative Commons Attribution (CC BY) license (http://creativecommons.org/licenses/by/4.0/).

Article

Composite Gel Polymer Electrolytes Based on Organo-Modified Nanoclays: Investigation on Lithium-Ion Transport and Mechanical Properties

Cataldo Simari *, Ernestino Lufrano, Luigi Coppola and Isabella Nicotera *

Department of Chemistry and Chemical Technologies, University of Calabria, 87036 Rende, CS, Italy; ernestino.lufrano@unical.it (E.L.); luigi.coppola@unical.it (L.C.)
* Correspondence: cataldo.simari@unical.it (C.S.); isabella.nicotera@unical.it (I.N.)

Received: 26 July 2018; Accepted: 21 August 2018; Published: 24 August 2018

Abstract: Composite gel polymer electrolytes (GPEs) based on organo-modified montmorillonite clays have been prepared and investigated. The organo-clay was prepared by intercalation of CTAB molecules in the interlamellar space of sodium smectite clay (SWy) through a cation-exchange reaction. This was used as nanoadditive in polyacrylonitrile/polyethylene-oxide blend polymer, lithium trifluoromethanesulphonate (LiTr) as salt and a mixture of ethylene carbonate/propylene carbonate as plasticizer. GPEs were widely characterized by DSC, SEM, and DMA, while the ion transport properties were investigated by AC impedance spectroscopy and multinuclear NMR spectroscopy. In particular, ^{7}Li and ^{19}F self-diffusion coefficients were measured by the pulse field gradient (PFG) method, and the spin-lattice relaxation times (T_1) by the inversion recovery sequence. A complete description of the ions dynamics in so complex systems was achieved, as well as the ion transport number and ionicity index were estimated, proving that the smectite clay surfaces are able to "solvatate" both lithium and triflate ions and to create a preferential pathway for ion conduction.

Keywords: gel polymer electrolytes; composites; montmorillonite clays; lithium batteries; PFG-NMR; self-diffusion coefficient; blend polymers

1. Introduction

Polymer electrolytes are regarded as one of the most promising candidates in advanced electrochemical applications, such as "smart" windows, displays, sensors, and more importantly, rechargeable lithium batteries [1–4]. For this last one, in particular, the research has focused for decades on gel-type membrane [5], generally achieved by immobilizing a liquid solution (for instance, a polar aprotic organic solvent or mixtures with a lithium salt) into a hosting polymeric matrix, such as poly(ethylene oxide) (PEO) and its derivatives (e.g., polyacrylonitrile (PAN), poly(vinylidene fluoride) (PVDF), poly(methyl methacrylate) (PMMA)) [6,7]. Respect to liquid electrolytes, in fact, gel polymer electrolytes (GPEs) are able to conjugate high ion conductivities with good mechanical strength, flexible geometry, reducing of liquid leaking and, thus, higher safety [8].

Owing to its ability to dissolve a large variety of salts, through interaction of its ether oxygen with cations, PEO has been one of the most extensively studied polymer used to prepare solid-state electrolytes, lighter, thinner, and safer for lithium-ion polymer batteries [9,10].

Thought, the low ionic conductivities at room temperature (10^{-6}–10^{-8} S cm^{-1}), the Li$^+$ transference number lower than 0.5 and the poor mechanical strength, still hinder the large scale diffusion of PEO-based device. Conversely, PAN ensures an ionic conductivity of circa 10^{-3} S cm^{-1}, satisfactory flame and mechanical resistances, but the dimensional stability of gels is poor [11,12]. After GPE preparation, in fact, a phase separation between the encapsulated electrolyte solution and the PAN matrix typically occurs, leading to a leakage problem and, thus, the passivation phenomena of the lithium electrode when in contact with the gel, as well as failure of the electrode/electrolyte contact both resulting in a dramatic reduction of the ionic conductivity.

One of the strategy undertaken to bypass the drawbacks is the blending method, according to which two or more polymers are mixed to obtain a blend electrolyte. As already probed [13–16] the method allows to easily control a large number of factors, directly affecting the thermal, mechanical and electrical properties of the final polymer electrolytes. By mixing PMMA and PVdF polymers, Nicotera and coworkers obtained a blend with remarkable improvement of mechanical stability respect to unblended polymers [17]. Helan et al. have been reported outstanding thermal stability up to 230 °C for PAN/PMMA blends, but with quite low ionic conductivity, of the order of 2×10^{-7} S cm^{-1} [18]. Very interesting electrical behavior and dimensional stability have been obtained by Choi et al. on PEO-PAN blend gel electrolytes, despite no evidence regarding mechanical resistance being provided [19].

An alternative approach for creating gel electrolyte system with improved mechanical properties and electrochemical performances foresees the incorporation of nanoscale organic/inorganic fillers within the polymer matrix [20]. The addition of SiO_2 [21], Al_2O_3 [22], TiO_2 [23], and other metal oxides [24,25] generally act as solid plasticizers, softening the polymer backbone and, thus, enhancing the segmental motion of the hosting polymer which, in turn, results in improved ion conductivity.

Among inorganic fillers, layered nanoparticles based on clays have been actively investigated lately since they offer a large number of interesting properties such as high cation exchange capacity, large chemically active surface area, outstanding swelling ability, intercalation, catalytic activity, and high chemical and thermal stability. Finally, the properties of the smectite nanoclays can be tailored using simple chemical methods such as intercalation with organic or inorganic guest molecules. From the above, the dispersion of proper clay minerals within the polymer matrix could enhance the ionic conductivity improving at the same time the strength and heat resistance of the GPE.

Smectite clay with different particle sizes has been effectively tested as filler for the preparation of PEO nanocomposite electrolytes, demonstrating a discrete improvement of ionic conduction [26]. Kurian et al. [27] have shown that the surface modification of clay by ion exchange reactions with cationic organic surfactants such as alkyl amines, enhance the chemical affinity with the polymer matrix, leading to exfoliation of the clay particles and improving the gel's strength. Organic montmorillonite (MMT) prepared by ion exchange with HTAB was dispersed in PAN polymer, obtaining a composite GPEs with improved thermal stability and ionic conductivity [28].

Despite the efforts, however, there is still the need to design a gel electrolyte able to guarantee adequate electrical performance without sacrificing mechanical strength and thermal resistance. In the present study, PAN/PEO blend (80:20 weight ratio) polymers were used in order to prepare nanocomposite GPEs with an organo-modified clay. Specifically, hydrated sodium calcium aluminum magnesium silicate hydroxide (SWy-2, Nanocor) was the natural montmorillonite/smectite clay selected since it is relatively inexpensive, widely available and has small particle size as well as it shows good intercalation capability. The organo-modification of the SWy-2 (org-SWy) was achieved by ion exchange reaction with hexadecyltrimethyl ammonium bromide (CTAB). The filler loading of org-SWy in the GPE was 10 wt % with respect to the polymers PAN/PEO. For the gel preparation, a mixture of ethylene carbonate (EC) and propylene carbonate (PC), with molar ratio EC:PC 1:0.4, was used as plasticizer, while lithium trifluoromethanesulfonate (LiTr) was the salt chosen.

In order to compare the effect of the clay on the gel properties, also not blended and filler-free GPE membranes were also prepared.

All the GPEs were investigated by thermal (DSC), morphological (scanning electronic microscopy-SEM) and mechanical (DMA) analysis, while the ion transport studies were conducted by electrochemical impedance spectroscopy (EIS) and by multinuclear NMR spectroscopy. In particular, the ^{1}H, ^{7}Li, and ^{19}F pulse-field-gradient (PFG) method was employed to obtain a direct measurement of the self-diffusion coefficients both of ions and solvents plasticizers (EC/PC), while the spin-lattice relaxation time (T_1) was obtained by the inversion recovery sequence.

The combination of the electrochemical and NMR data has provided a wide description of the ions dynamics inside the so complex systems, as well as information on ion associations and interactions between polymers, filler and ions.

2. Materials and Methods

2.1. Materials

Poly(ethylene oxide) (PEO, M.W. 5,000,000), polyacrylonitrile (PAN), lithium trifluoromethanesulfonate (LiCF$_3$SO$_3$ or LiTr, 99.95%), ethylene carbonate (EC, 98%), propylene carbonate anhydrous (PC, 99.7%), and hexadecyltrimethyl ammonium bromide (CTAB, 98%) were purchased from Sigma Aldrich, Milan, Italy and used as received.

Natural smectite Wyoming montmorillonite (SWy-2) has obtained from the Source Clay Minerals Repository, University of Missouri Columbia, MO, USA. The cation exchange capacity (CEC), measured by the Co(II) procedure, is equal to 80 mequiv. per 100 g of clay, charge density 0.6 e^{-1}/unit cell (the unit cell is the Si$_8$O$_{20}$ unit) and particle size around 200 nm. The structural formula is Na$_{0.62}$[Al$_{3.01}$Fe(III)$_{0.41}$Mg$_{0.54}$Mn$_{0.01}$Ti$_{0.02}$](Si$_{7.98}$Al$_{0.02}$) O$_{20}$(OH)$_4$.

2.2. Synthesys of Organo-Modified Clay (Org-Swy)

SWy-2 were first fractioned to <2 µm by gravity sedimentation and purified by well-established procedures in clay science [29]. For the chemical modification, the cation exchange capacity of smectite clay has been exploited. CTAB (0.4 g) was dissolved in boiling deionized water until complete dissolution, then the resulting solution has been dropwise added, under vigorous stirring, to a dispersion of SWy-2 (1.0 g) in deionized water at 60 °C and left for 6 h to achieve the total cationic exchange. Finally, the mixture solution was separated by centrifugation, rinsed repeatedly with deionized water until Br− was completely removed, and dried for 24 h at 90 °C.

2.3. GPE Membrane Preparation

The solvent casting technique has been used to prepare both blended and not blended membranes, by immobilization of a lithium salt solution in a polymer matrix.

The required amounts of PAN and PAN-PEO (80/20 blend ratio) were dissolved in anhydrous dimetylformammide (DMF). The solution was stirred for several hours at 60 °C, until a homogeneous mixture was obtained and, after complete dissolution, the electrolyte solution was added. For the electrolyte solution, LiCF$_3$SO$_3$ was dissolved in a mixture of EC and PC with a fixed molar ratio (1:0.4). The lithium content, expressed as the ratio between the number of EC-PC moles and the LiTr moles (also O/Li ratio), was 10/1. Finally, the polymers/plasticizers [PAN:(EC-PC) and (PAN-PEO)/(EC-PC)] weight ratio was of 26:74.

For the nanocomposite GPEs, the appropriate amount of organo-modified clay has been added to DMF, mechanically stirred for 16 h and sonicated for 8 h to obtain a homogeneous dispersion. The dispersion was then added dropwise to the polymer solution, followed by further sonication and stirring. Here composite membranes with 10% of filler loading with respect to the polymer were prepared. The membranes were achieved by casting the solution on the aluminum plate at 50 °C overnight to favor the evaporation of DMF.

2.4. Characterization Techniques

NMR measurements were performed with a BRUKER AVANCE 300 Wide Bore spectrometer working at 116.6 MHz on ^{7}Li, and 282.4 MHz on ^{19}F, respectively. The employed probe was a Diff30 Z-diffusion 30 G/cm/A multinuclear with substitutable RF inserts. Spectra were obtained by transforming the resulting free-induction decay (FID) of single π/2 pulse sequences.

The pulsed field gradient stimulated-echo (PFG-STE) method [30] was used to measure the self-diffusion coefficients of lithium and triflate ions. The sequence consists of three 90° RF pulses

($\pi/2 - \tau_1 - \pi/2 - \tau_m - \pi/2$) and two gradient pulses that are applied after the first and the third RF pulses, respectively. The echo is found at time $\tau = 2\tau_1 + \tau_m$. Following the usual notation, the magnetic field pulses have magnitude g, duration δ, and time delay Δ. The FT echo decays were analyzed by means of the relevant Stejskal–Tanner expression:

$$I = I_0 e^{-\beta D}$$

Here I and I_0 represent the intensity/area of a selected resonance peak in the presence and in absence of gradients, respectively. β is the field gradient parameter, defined as $\beta = [(\gamma g \delta)]^2 (\Delta - \delta/3)$; D is the measured self-diffusion coefficient.

In these experiments, the used experimental parameters were: δ = 3 ms, time delay Δ = 30 ms, and the gradient amplitude varied from 350 to 1000 G cm^{-1}. Based on the very low standard deviation of the fitting curve and repeatability of the measurements, the uncertainties in D values are estimated to about 3%.

Finally, longitudinal relaxation times (T_1) of ^{7}Li and ^{19}F were measured by the inversion-recovery sequence ($\pi - \tau - \pi/2$). All the NMR measurements were run by increasing temperature step by step from 20 to 80 °C, with steps of 10 °C, and leaving the sample to equilibrate for about 20 min at each temperature value.

From D_{Li} and D_F self-diffusion coefficients, σ_{NMR} values were calculated according with the Nernst-Einstein equation:

$$\sigma_{NMR} = \frac{F^2 c_{LiTr}}{RT}(D_{Li^+} + D_{F^+})$$

Here, F is the Faraday constant, R is the molar gas constant, T is the temperature to which D has been measured and c_{LiTr} is the salt concentration.

The ionic conductivity (σ, S cm^{-1}) was measured by impedance spectroscopy recorded at OCV with an oscillating potential of 10 mV in the frequency range 0.1–1 $\times$ 10^6 Hz using a PGSTAT 30 (MetrohmAutolab) potentiostat/galvanostat/FRA. GPEs were sandwiched between two disks of conductive carbon cloth, placed between two stainless steel electrodes and assembled in a homemade two-electrode cell. The impedance responses of the cell were analyzed using MetrohmAutolab NOVA software and the bulk resistance (R_b) was extracted from the intercept of the low frequency signal in the Nyquist plot. The equation for calculating the conductivity is:

$$\sigma = \frac{l}{R_b * A}$$

where l is the thickness of the membrane and A is the area of the carbon cloth electrode.

Dynamic mechanical analysis (DMA) measurements were carried out on a Metravib DMA/25 analyzer equipped with a shear jaw for film clamping. Frequency sweep experiments were collected by subjecting a rectangular film to a dynamic strain of amplitude 10^{-4} in the range between 0.2 and 20 Hz. For temperature sweep (time cure) experiments a dynamic strain of amplitude 10^{-4} at 1 Hz was applied from 25 to 160 °C with a heating rate of 2 °C/min. A periodic sinusoidal displacement was applied to the sample, and the resultant force was measured. The damping factor, tan d, is defined as the ratio of loss (E") to storage (E') modulus.

The thermal behaviors were investigated by Setaram 131 DSC. Samples were hermetically sealed and cooled from room temperature to -40 °C using liquid N$_2$. Measurements were carried out from -30 °C up to 120 °C at the scan rate of 10 °C/min and purging nitrogen gas.

Finally, the membrane's morphology was investigated by scanning electron microscopy (SEM, Cambridge Stereoscan 360, Santa Clara, CA, USA). To observe the membrane cross-sections, the samples were first frozen and fractured in liquid nitrogen, to guarantee a sharp fracture without modifications of the morphology, and then observed with SEM. The samples were sputter-coated with a thin gold film prior to SEM observation.

3. Results and Discussion

3.1. Morphological, Thermal, and Mechanical Characterizzation of the GPEs

The organo-modification of the clay's layers has as the main objective of favoring a good and homogeneous dispersion of the nanoparticles into the hosting matrix. For this purpose, hexadecyltrimethyl ammonium bromide was used as organophilic reagents: the quaternary ammonium group should allow an easy intercalation into the hydrophilic clay layers while the long alkyl chain should enhance the affinity between particles and polymer chains [31]. The photos of the four gel electrolytes prepared in this study are reported in Figure 1. They all appear opalescent, while the introduction of the org-SWy causes a slight yellowing of the resulting GPEs (Figure 1b,d). However, they are very dense and homogeneous, and there is no evidence of phase segregation between PAN and PEO polymers into the blended gels, indicating that the proposed method allows to obtain a homogenous and stable mixtures of polymers. Further, no clay particles crystals were observed, confirming that the chemical modification of the layers' surface improves the clay/polymer interaction and, thus, highly homogeneous composite membranes, without formation of agglomerates or clusters, can be prepared.

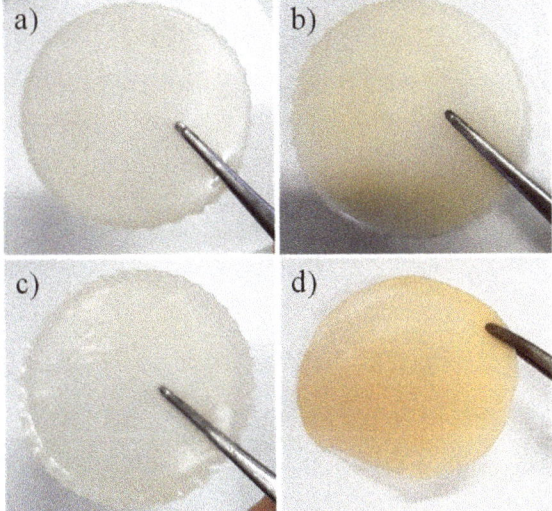

Figure 1. Photos of the prepared GPEs based on: (**a**) PAN, (**b**) PAN/org-SW, (**c**) PAN:PEO blend, and (**d**) PAN:PEO/org-SWy.

Scanning electron microscopy (SEM) coupled with BSE (backscattered electrons) was used to deeper investigate the morphology of the composite membranes. The BSE technique is generally used to detect contrast between areas with different chemical compositions (elements with high atomic number backscatter electrons more efficiently than light elements, appearing brighter in the image). By comparing the SEM-BSE images obtained on pristine PAN and PAN/org-SWy electrolytes, shown in Figure 2a,b, respectively, it clearly emerges that the presence of the filler particles severely affect the film morphology. The porous structure of the PAN based gel disappear in the composite gel, becoming a very dense membrane. Sporadic particle aggregations are also detectable, as expected if we take into account the large percentage of filler added into the polymer matrix (10 wt %). However, the average particles size of such aggregates is circa 500 nm, therefore, it can be stated that the nano-sized and homogeneous dispersion of clay layers was achieved in these composite GPEs. Concerning the blends

(Figure 2c,d), SEM + BSE images give clear evidence that no phase separation occurs between the two polymers, as well as the presence of PEO allows the reduction of the number of nanosized aggregates in the composite blend electrolyte by virtue of a greater affinity between poly(ethylene oxide) chains and the org-SWy lamellae.

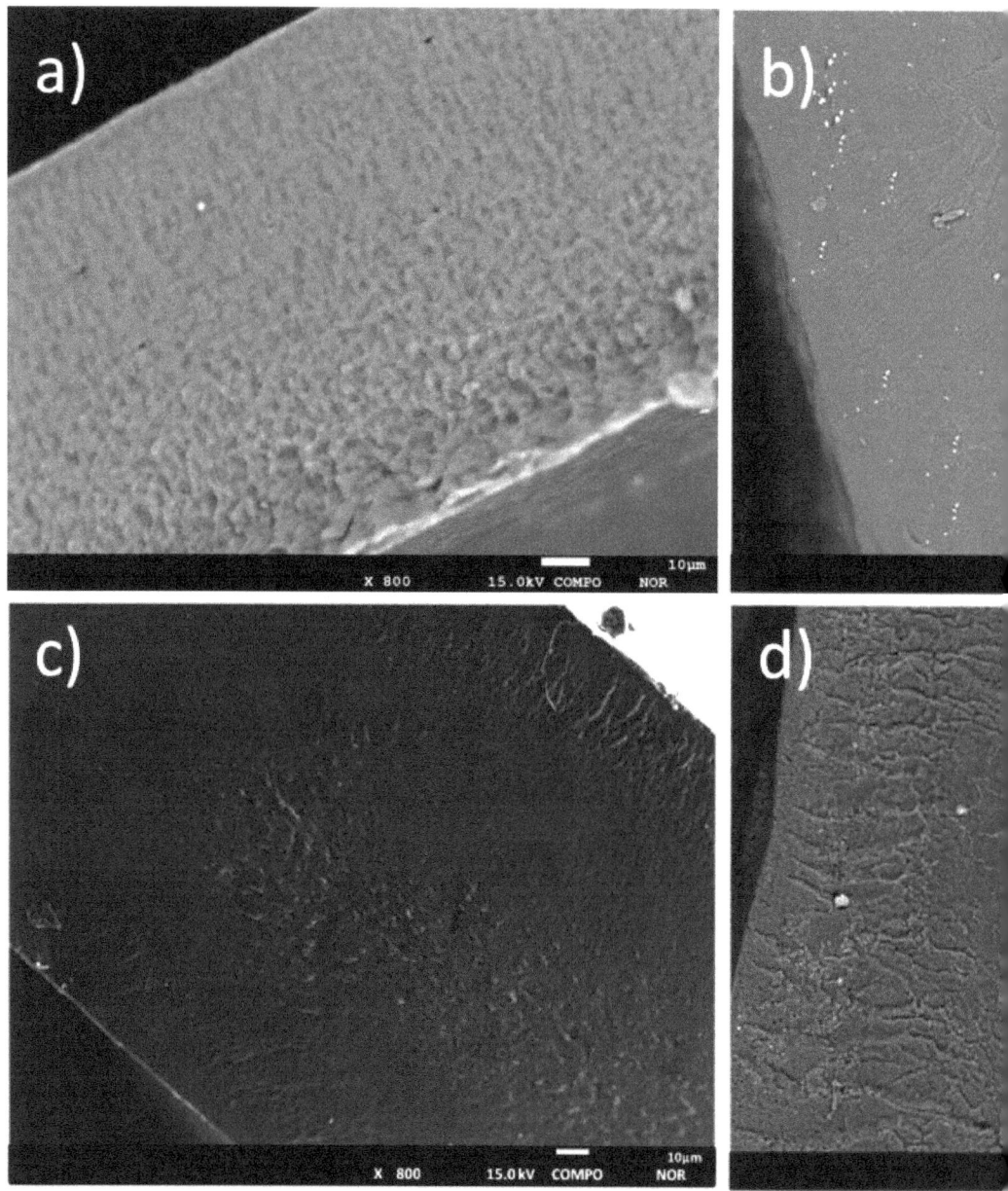

Figure 2. Cross-sectional SEM + BSE images of the GPEs based on: (**a**) PAN; (**b**) PAN/org-SWy; (**c**) PAN:PEO; and (**d**) PAN:PEO/org-SWy.

The analysis of the thermal properties of the prepared electrolytes has been carried out by DSC, and the thermograms collected in the temperature range between −30 and 120 °C are showed in Figure 3. For clarity it must be noticed that, in order to highlight the peaks, an enlarged scale was used.

The PAN-based gel shows two endothermic peak, the first one narrow, at about 71 °C (T_{gI}), and the second broad peak at circa 100 °C (T_{gII}). It was already demonstrated [32] that unoriented PAN has a "two-phase" morphology consisting of laterally-ordered and amorphous domains, both in a glassy state at room temperature, thus leading to two glass transition at 100 and 150 °C, respectively. In our films, the inclusion of EC/PC plasticizer lowers both T_g respect to pristine PAN, as a consequence of the reduction of the crystallites size [33]. The dispersion of org-SWy platelets leads to a large shift of the transitions of both laterally-ordered and amorphous domains (red line in the Figure 3), and also to a reduction of the peaks intensities, suggesting interactions between the organo-modified silicate layers and the polymer chains. It can be hypothesized that org-SWy particles increase the distance between polymer chains and, hence, diminish their capability to re-aggregate in glassy domains.

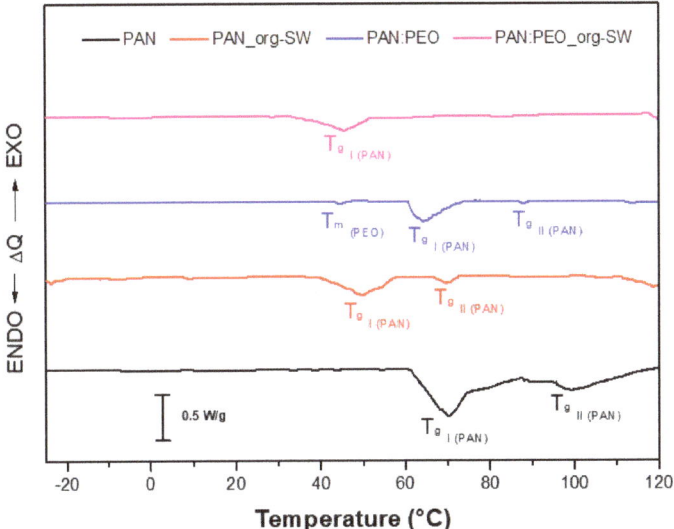

Figure 3. DCS thermograms of the GPEs membranes in the temperature range −30 °C up to 120 °C, with a scan rate of 10 °C min^{-1}.

Focusing on blended electrolytes, in PAN:PEO films a small peak at 44 °C appears, corresponding to the typical temperature at which PEO crystalline domains becomes rubbery amorphous phase (T_m PEO). Finally, the nanocomposite blend electrolyte shows a single broad peak at 45 °C ascribed to the T_{gI} of PAN while disappear the T_m of PEO. The result can be explained in terms of larger chemical affinity between clay platelets and PEO chains, which reduces PEO re-crystallization and, at the same time, favors the dispersion of filler's particles within the polymer matrix.

Concerning the mechanical properties of the GPEs systems, the measurements were performed by dynamic mechanical analysis, by using a shear jaw for films sample holder. It is worth pointing out that, generally, oscillatory rheological tests on typical GPEs are carried out by using a plate-plate geometry, while, in this case, due to the solid-like nature of our gels, a typical DMA configuration for thin films was used.

Figure 4a shows the storage modulus (E') in the frequency range of 0.2–20 Hz measured at 25 °C: Except for one sample which will be discussed later, E' shows values above 10^7 Pa, significantly higher than other gels reported in the literature [34–36], and it reaches 10^8 Pa upon inclusion of org-SWy lamellae in the PAN matrix, indicating an increase in the rigidity of the system. Blending PAN and PEO polymers also results in an enhanced storage modulus as a consequence of the increased overall crystallinity of the polymeric matrix. However, completely unexpected is the net reduction of the storage modulus of the composite blend PAN:PEO/org-SWy electrolyte to 10^6 Pa. This evidence can be explained by taking into account the DSC data seen above. The inclusion of the clay into the polymer matrices prevents the reorganization of PAN and PEO chains into crystalline stacks, affecting the mechanical strength of the film but, at the same time, improves the flexibility of polymer chains, with important implications on the transport properties of this electrolyte gel. However, the temperature-sweep test shown in Figure 4b demonstrates that this composite still maintains the typical strong-gel behavior, likely due to the interactions between clay platelets and polymer chains. In fact, at least up to 160 °C, the storage modulus E' exceeds significantly the loss modulus E'', indicating that the gel responds elastically at small deformations and its microstructure is unchanged over this temperature range. The slight slope of the moduli is indicative of an evolution towards a "weak-gel" configuration, nonetheless, no crossover between the moduli occurs; therefore, the structure of the gel is preserved.

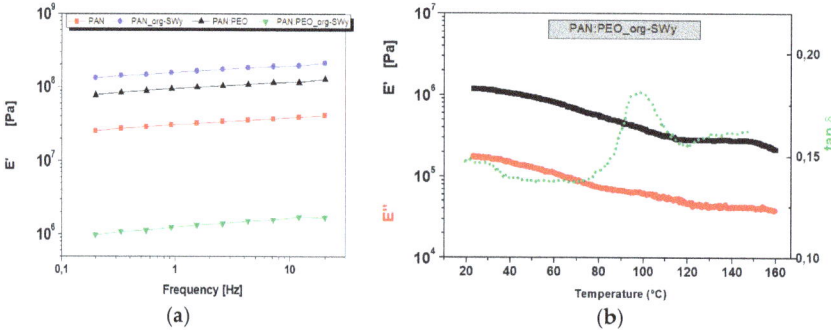

Figure 4. Frequency sweep at 25 °C of the different GPEs (**a**); and the temperature sweep test, from 20 °C to 160 °C for PAN:PEO/org-SWy electrolyte (**b**).

3.2. Transport Properties of Ions

The ionic conductivities of the prepared gel polymer electrolytes were investigated by EIS analysis. The impedance Nyquist plots of two representative GPEs are reported in Figure 5. The insets in each graph show an enlargement of the low resistance region, where the semicircle is achieved. In fact, the spectra show two well-defined regions: a semicircular region at high frequency range (attributed to ion conduction process in the bulk of the gel polymer electrolyte) followed by a straight line inclined at constant angle of circa 40° to the real axis at low frequency range related to the effect of blocking electrodes [37,38]. By comparing the spectra of PAN gels (Figure 5a) and of PAN/org-SWy nanocomposite (Figure 5b), we can notice that the semicircle of the nanocomposite appears as depressed, i.e., it is not completed in the frequency range used, although very high (1 MHz). This indicates that multiple processes and/or mechanisms of conduction simultaneously coexist [27]. A similar trend has been also observed in blended PAN:PEO/org-SWy electrolyte, even if less pronounced.

From the fitting of the semicircle in the high-frequency region, the electrolyte resistance was estimated and the ionic conductivity (σ) calculated according to the formula reported in the experimental and displayed in Figure 6. It clearly emerges that PAN-SWy nanocomposite gel is the less conductive electrolyte. Such an outcome can be explained by considering the changing of the gel morphology upon addition of the clay to the polymer matrix, as discussed above, which becomes dense, as well as more rigid (higher Young's modulus). Therefore, the polymer chains experience lower flexibility, as well as a large reduction of liquid electrolyte mobility is expected by the decrease of the membrane porosity, both contributing to the reduction of the ion conduction.

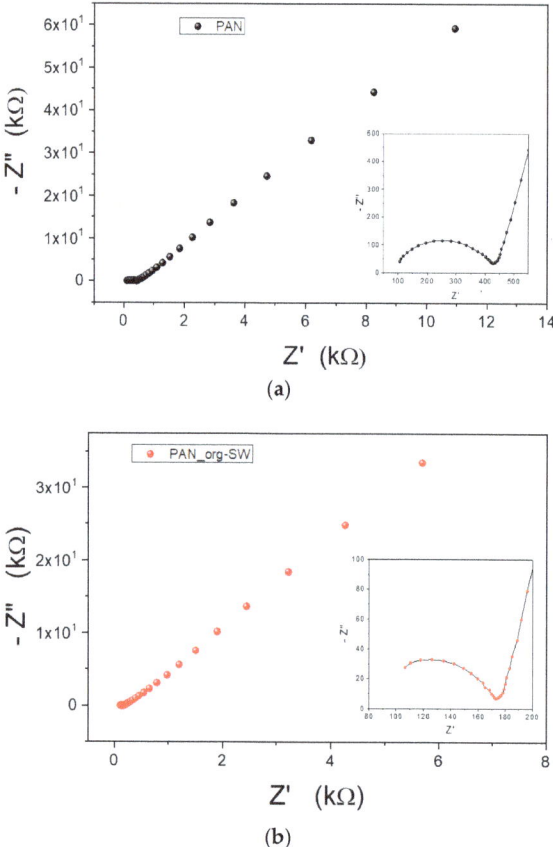

Figure 5. Nyquist plots of the impedance measured for PAN gel (**a**) and PAN/org-SWy nanocomposite gel (**b**).

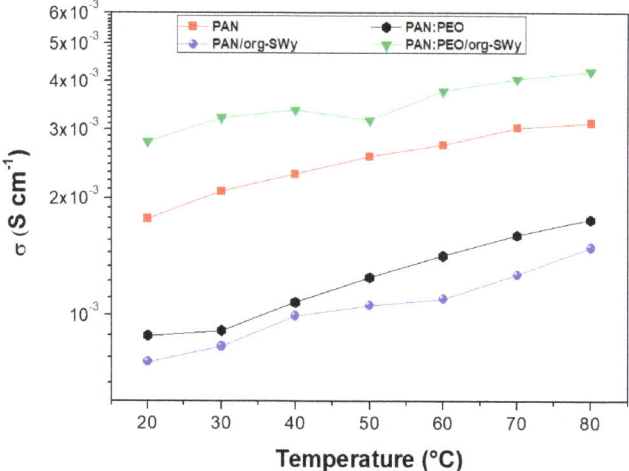

Figure 6. Temperature dependence of ionic conductivity for the gel polymer electrolytes investigated.

Similar discussion can be made on the PAN:PEO blend gel, where the enhanced membrane rigidity caused by the increased number of crystalline domains of PEO significantly affects σ compared to the unblended PAN.

The best result was achieved by the addition of 10 wt % of organo-modified SWy in the PAN:PEO blend, which displays the highest ion conductivity over the whole temperature range, with a σ of almost 2.8 mS/cm at r.t. Comparing to similar GPEs reported in the literature, these conductivities are surely remarkable: e.g., they are two orders of magnitude higher than hybrid electrolytes composed of PEO and glass-ceramic particles (2.81×10^{-2} mS/cm) [26] and three orders higher than PEO containing conductive microsized particles (1×10^{-3} mS/cm) [39], while they are close to those reported by He et al. [31] for a PAN/organic montmorillonite system (2.23 mS/cm), even if, here, an electrolyte uptake of ca. 300% was needed, resulting in deterioration of the membrane stability. Accordingly, it can be stated that the PAN:PEO composite gels are able to guarantee good polymer chain flexibility together with outstanding mechanical and thermal resistance, making these systems particularly attractive as solid electrolytes for lithium batteries.

It is well known that the ionic conductivity obtained by EIS only refers to the mobility of charged species, with no possibility to distinguish between the cation and the anion. Conversely, NMR methods allow to discriminate and selectively investigate the mobility of Li^+ and the corresponding counterion, confirming the effectiveness regarding the investigation of ions dynamics inside the complex systems, as well as information on ion associations and interactions between polymers, filler, and ions. Accordingly, NMR was used here to investigate the transport properties of both lithium cations and triflate anions, by detecting the ^7Li and ^{19}F spin-nuclei, respectively.

Figure 7 displays the lithium self-diffusion coefficients (D_{Li}) measured on the GPEs' membranes, both unblended (left) and blended (right), respectively. In agreement with the conductivity seen above, the addition of org-SWy to PAN reduces the lithium mobility while it has beneficial impact in the PAN:PEO blend. However, very interesting is the bi-exponential decay of the echo-signal obtained in both composite systems, observed also for the D_F (diffusion values for ^{19}F, not reported in the graph). This result indicates that two different mechanism for the diffusing species coexist as a consequence of the presence of the clay lamellae. The aluminosilicate platelets possess a fixed negative charge and the quaternary ammonium group of CTAB molecules was chosen as intercalating cation. Ions are solvated both from the clay layers ("lamellae-solvation") and from the EC/PC solvents ("bulk-solvation") and, of course, the polymers play their role in such coordination.

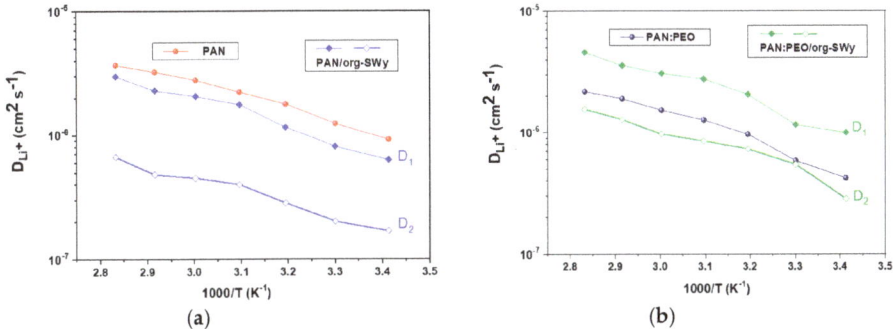

Figure 7. Arrhenius plots of ^7Li self-diffusion coefficients from 20 to 80 °C measured on PAN-based electrolytes (**a**) and blended systems (**b**).

Ions involved in the "bulk-solvation" show higher mobility (D_1) respect to that one involved in the "lamellae-solvation" (D_2).

Such a hypothesis was confirmed by the spin-lattice relaxation time (T_1), which, compared to diffusion, reflects more localized motions, including both translation and rotation on a time scale comparable to the reciprocal of the NMR angular frequency (few nanoseconds). T_1 quantifies the energy transfer rate from the nuclear spin system to the neighboring molecules (the lattice). The stronger the interaction, the quicker the relaxation (shorter T_1). Figure 8 reports the Arrhenius plots of T_1 measured on the different GPEs for ^{7}Li and ^{19}F, respectively. It is clear that the introduction of org-SWy particles produces a decrease of T_1, both for ^{7}Li and ^{19}F. This outcome can be ascribed to the stronger overall interactions of the ions with the lattice, i.e., lithium ions interact with negative charged surface of the platelets, while counterions solvate the quaternary ammonium groups of the organo-surfactant. In other words, ions experience a lower degree of freedom resulting in shorter T_1 values.

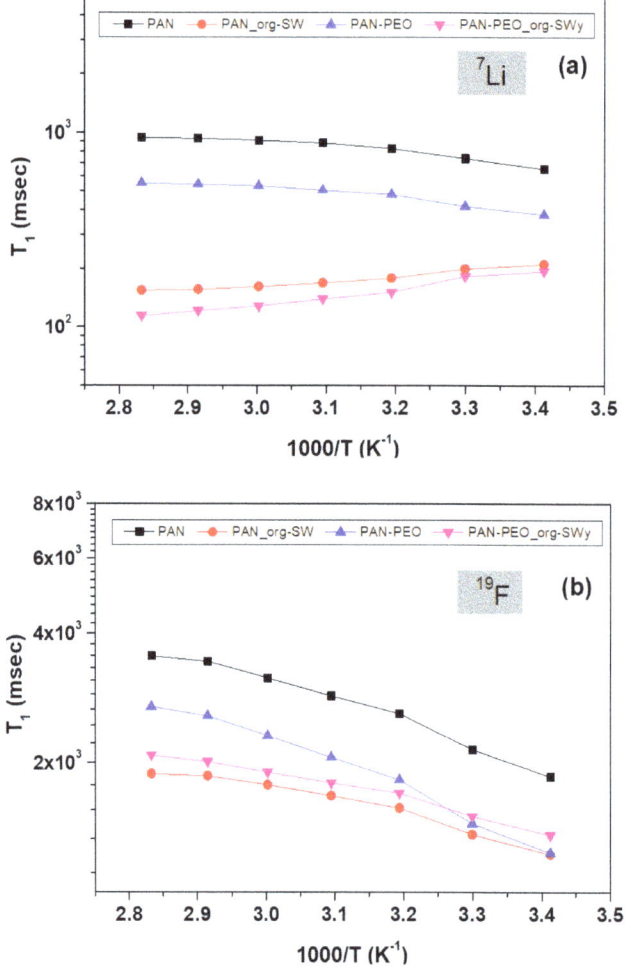

Figure 8. Arrhenius plot of ^{7}Li (**a**) and ^{19}F (**b**) spin-lattice relaxation time from 20 °C up to 80 °C.

According to the Nernst-Einstein equation, conductivity values (σ_{NMR}) were calculated from D_{Li} and D_F for the different GPEs and compared with the experimental ion conductivity (σ_{EIS}) in Table 1 (for two representative gels). We need to consider that differently from σ_{EIS}, σ_{NMR} is affected not from the mobility of all species containing ^{7}Li and ^{19}F, including neutral ion pairs, and not only from the charged species. Therefore, it is not unusual for the NMR conductivity to be greater than the experimental σ, in particular when ion associations occurs. By considering the bi-exponentiality of both Li$^+$ and F$^-$ diffusion, and based on the hypothesis discussed above, we managed to calculate an average of D_1 and D_2 weighed with respect to the amount of filler added, i.e., 10 wt %. It is evident from the data reported that NMR conductivity values are always much higher than experimental ones suggesting the presence of a large number of ion pairing. This is also confirmed by the ionicity indices reported in Table 1 and computed as the ratio $\sigma_{EIS}/\sigma_{NMR}$.

Table 1. Comparison between σ_{EIS} and σ_{NMR} (in Ms cm^{-1}), ionicity index and lithium transport number for PAN and PAN:PEO/org-SWy electrolytes.

T (°C)	PAN				PAN-PEO + 10% SW			
	σ_{EIS}	σ_{NMR}	Ionicity	t_{Li+}	σ_{EIS}	σ_{NMR}	Ionicity	t_{Li+}
20	1.77	3.92	0.45	0.40	2.79	4.31	0.68	0.68
30	2.08	5.02	0.41	0.41	3.22	4.74	0.65	0.67
40	2.31	6.96	0.33	0.41	3.37	7.84	0.53	0.59
50	2.56	8.19	0.31	0.43	3.17	9.07	0.45	0.56
60	2.74	9.68	0.28	0.44	3.77	9.82	0.48	0.56
70	3.03	11.1	0.27	0.43	4.05	11.20	0.42	0.57
80	3.12	11.8	0.27	0.45	4.24	12.33	0.38	0.58

PAN gel, our reference's system, shows an ionicity close to 0.45. This suggests that 55% of Li$^+$ and Tr$^-$ exist as neutral ion pairs, which is typical for GPEs. The addition of filler particles into the blend increases the level of salt dissociation, likely due to the high dielectric constants of the charged organo-modified smectite clays that should also help to prevent the ionic association. Both phenomena leads to an ionicity index of 0.68 at r.t., which is a particularly high value for a double-ion solid-state electrolyte. Ionic association increases by increasing the temperature [11,39], therefore, the ionicity index decreases.

Finally, an important parameter for allowing a proper operation of the polymer electrolyte in real device is the lithium transport number (t_{Li+}). It was calculated in this work according to the following equation and reported in Table 1:

$$t_{Li+} = \frac{D_{Li+}}{D_{Li+} + D_{F-}}$$

The PAN:PEO/org-SWy electrolyte shows a value of 0.68 at r.t., much higher than the PAN-gel and also the typical GPEs, for which values lower than 0.30 are generally reported [26,40–43]. GPEs with higher lithium transport number, i.e., ca. 0.55, has also been reported, but the ion conductivities are quite low [44]. The reasons of the improved t_{Li+} in our blend composite membrane can be multiple and synergistic: (i) the organo-clay particles have a plasticizing effect, lowering the cristallinity and, thus, improving the flexibility of polymer chains, favoring the Li$^+$ transport through polymer segmental motions; and (ii) electrostatic interactions between the filler surface and lithium can create a preferential pathways for lithium conduction.

4. Conclusions

Organo-modified smectite clay particles were prepared and dispersed into PAN and PAN:PEO blend polymers in order to prepare hybrid gel polymer electrolytes. Morphological studies proved that the procedure herein proposed allows to avoid phase separation between PAN and PEO as well as guarantee high nano-dispersion of the clay particles in the polymer matrix. The presence of the clay platelets strongly affected morphology, thermal and mechanical stability and electrochemical properties of the GPEs. In particular, outstanding behavior was displayed by the PAN:PEO/org-SWy membrane. ^{7}Li and ^{19}F NMR spectroscopy was successfully applied to get a complete description of the ions dynamics in so complex systems, probing as the smectite clay surfaces are able to "solvate" both lithium and triflate ions, preventing the ion pairing (as also confirmed by the high ionicity index) and creating preferential pathways for lithium conduction.

Author Contributions: Conceptualization, I.N. and C.S.; Methodology, C.S., E.L. and L.C.; Validation, I.N.; Investigation, C.S. and E.L.; Resources, I.N.; Data Curation, C.S. and L.C.; Writing-Original Draft Preparation, C.S.; Writing-Review & Editing, I.N.; Supervision, I.N.; Project Administration, I.N.; Funding Acquisition, I.N.

Funding: This work was supported by the European Community's Seventh Framework Program (FP7 2007-2013) through the MATERIA Project (PONa3_00370).

Acknowledgments: The authors would like to thank Mariano Davoli, University of Calabria, for his precious support in the morphological characterization of the membranes.

Conflicts of Interest: The authors declare no conflict of interest.

References

1. Scrosati, B.; Neat, R.J. *Applications of Electroactive Polymers*; Scrosati, B., Ed.; Chapman & Hall: London, UK, 1993; Volume 6.
2. Ngai, K.S.; Ramesh, S.; Ramesh, K.; Juan, J.C. A review of polymer electrolytes: Fundamental, approaches and applications. *Ionics* **2016**, *22*, 1259–1279. [CrossRef]
3. Cossari, P.; Simari, C.; Cannavale, A.; Gigli, G.; Nicotera, I. Advanced processing and characterization of Nafion electrolyte films for solid-state electrochromic devices fabricated at room temperature on single substrate. *Solid State Ion.* **2018**, *317*, 46–52. [CrossRef]
4. Shujahadeen, B.A.; Thompson, J.W.; Kadir, M.F.Z.; Hameed, M.A. A conceptual review on polymer electrolytes and ion transport models. *J. Sci.: Adv. Mater. Devices* **2018**, *3*, 1–17.
5. Song, J.Y.; Wang, Y.Y.; Wan, C.C. Review of gel-type polymer electrolytes for lithium-ion batteries. *J. Power Sources* **1999**, *77*, 183–197. [CrossRef]
6. Xue, Z.; He, D.; Xie, X. Poly(ethylene oxide)-based electrolytes for lithium-ion batteries. *J. Mater. Chem. A.* **2015**, *3*, 19218–19253. [CrossRef]
7. Nicotera, I.; Coppola, L.; Oliviero, C.; Ranieri, G.A. Properties and Impedance Spectroscopy of PMMA-PVdF Blend and PMMA Gel Polymer Electrolytes for Advanced Lithium Batteries. *Ionics* **2005**, *11*, 87–94. [CrossRef]
8. Gwon, H.; Hong, J.; Kim, H.; Seo, D.H.; Jeon, S.; Kang, K. Recent progress on flexible lithium rechargeable batteries. *Energy Environ. Sci.* **2014**, *7*, 538–551. [CrossRef]
9. Li, W.; Pang, Y.; Liu, J.; Liu, G.; Wang, Y.; Xia, Y. A PEO-based gel polymer electrolyte for lithium ion batteries. *RSC Adv.* **2017**, *7*, 23494–23501. [CrossRef]
10. Polu, A.R.; Rhee, H.W. Ionic liquid doped PEO-based solid polymer electrolytes for lithium-ion polymer batteries. *Int. J. Hydrogen Energy* **2017**, *42*, 7212–7219. [CrossRef]
11. Nicotera, I.; Oliviero, C.; Ranieri, G.A.; Spadafora, A.; Castriota, M.; Cazzanelli, E. Temperature evolution of thermoreversible polymer gel electrolytes LiClO4/ethylene carbonate/polyacrylonitrile. *J. Phys. Chem.* **2002**, *117*, 7373. [CrossRef]
12. Yue, L.; Ma, J.; Zhang, J.; Zhao, J.; Dong, S.; Liu, Z.; Cui, G.; Chen, L. All solid-state polymer electrolytes for high-performance lithium ion batteries. *Energy Storage Mater.* **2016**, *5*, 139–164. [CrossRef]
13. Manuel Stephan, A.; Nahm, K.S. Review on composite polymer electrolytes for lithium batteries. *Polymer* **2006**, *47*, 5952–5964. [CrossRef]
14. Xi, J.Y.; Qiu, X.P.; Li, J.; Tang, X.Z.; Zhu, W.T.; Chen, L.Q. PVDF-PEO blends based microporous polymer electrolyte: Effect of PEO on pore configurations and ionic conductivity. *J. Power Sources* **2006**, *157*, 501–506. [CrossRef]
15. Tao, C.; Gao, M.H.; Yin, B.H.; Li, B.; Huang, Y.P.; Xu, G.W.; Bao, J.J. A promising TPU/PEO blend polymer electrolyte for all-solid-state lithium ion batteries. *Electrochim. Acta* **2017**, *257*, 31–39. [CrossRef]
16. Nunes, P.J.; Costa, C.M.; Lanceros, M.S. Polymer composites and blends for battery separators: State of the art, challenges and future trends. *J. Power Sources* **2015**, *281*, 378–398. [CrossRef]
17. Nicotera, I.; Coppola, L.; Oliviero, C.; Castriota, M.; Cazzanelli, E. Investigation of ionic conduction and mechanical properties of PMMA–PVdF blend-based polymer electrolytes. *Solid State Ion.* **2006**, *177*, 581–588. [CrossRef]

18. Helan Flora, X.; Ulaganathan, M.; Shanker Babu, R.; Rajendran, R. Evaluation of lithium ion conduction in PAN/PMMA-based polymer blend electrolytes for Li-ion battery applications. *Ionics* **2012**, *18*, 731–736. [CrossRef]
19. Choi, B.K.; Kim, Y.W.; Shin, H.K. Ionic conduction in PEO–PAN blend polymer electrolytes. *Electrochimica Acta* **2000**, *45*, 1371–1374. [CrossRef]
20. De Souza, F.L.; Leite, E.R. Hybrid polymer electrolytes for electrochemical devices. *Polym. Electrolytes Fundam. Appl.* **2010**, 583–602.
21. Wetjen, M.; Navarra, M.A.; Panero, S.; Passerini, S.; Scrosati, B.; Hassoun, J. Composite Poly(ethylene oxide) Electrolytes Plasticized by *N*-Alkyl-*N*-butylpyrrolidinium Bis(trifluoromethanesulfonyl)imide for Lithium Batteries. *ChemSusChem* **2013**, *6*, 1037–1043. [CrossRef] [PubMed]
22. Masoud, E.M.; Bellihi, E.A.A.; Bayoumy, W.A.; Mousa, M.A. Organic–inorganic composite polymer electrolyte based on PEO–LiClO$_4$ and nano-Al$_2$O$_3$ filler for lithium polymer batteries: Dielectric and transport properties. *J. Alloys Compd.* **2013**, *575*, 223–228. [CrossRef]
23. Plylahan, N.; Letiche, M.; Barr, M.K.S.; Djenizian, T. All-solid-state lithium-ion batteries based on self-supported titania nanotubes. *Electrochem. Commun.* **2014**, *43*, 121–124. [CrossRef]
24. Xiong, H.M.; Wang, Z.D.; Liu, D.P.; Chen, J.S.; Wang, Y.G.; Xia, Y.Y. Bonding polyether onto ZnO nanoparticles: An effective method for preparing polymer nanocomposites with tunable luminescence and stable conductivity. *Adv. Funct. Mater.* **2005**, *15*, 1751–1756. [CrossRef]
25. Panero, S.; Scrosati, B.; Sumathipala, H.H.; Wieczorek, W. Dual-composite polymer electrolytes with enhanced transport properties. *J. Power Sources* **2007**, *167*, 510–514. [CrossRef]
26. Zhao, Y.; Huang, Z.; Chen, S.; Chen, B.; Yang, J.; Zhang, Q.; Ding, F.; Chen, Y.; Xiaoxiong, X. A promising PEO/LAGP hybrid electrolyte prepared by a simple method for all-solid-state lithium batteries. *Solid State Ion.* **2016**, *295*, 65–71. [CrossRef]
27. Kurian, M.; Galvin, M.E.; Trapa, P.E.; Sadoway, D.R.; Mayes, A.M. Single-ion conducting polymer–silicate nanocomposite electrolytes for lithium battery applications. *Electrochim. Acta* **2005**, *50*, 2125–2134. [CrossRef]
28. Chen, Y.; Chen, Y.T.; Chen, H.C.; Lin, W.T.; Tsai, C.H. Effect of the addition of hydrophobic clay on the electrochemical property of polyacrylonitrile/LiClO$_4$ polymer electrolytes for lithium battery. *Polymer* **2009**, *50*, 2856–2862. [CrossRef]
29. Nicotera, I.; Enotiadis, A.; Angjeli, K.; Coppola, L.; Gournis, D. Evaluation of smectite clays as nanofillers for the synthesis of nanocomposite polymer electrolytes for fuel cell applications. *Int. J. Hydrogen Energy* **2012**, *37*, 6236–6245. [CrossRef]
30. Tanner, J.E. Use of the stimulated echo in NMR diffusion studies. *J. Chem. Phys.* **1970**, *52*, 2523–2526. [CrossRef]
31. He, C.; Liu, J.; Cui, J.; Li, J.; Wu, X. A gel polymer electrolyte based on Polyacrylonitrile/organic montmorillonite membrane exhibiting dense structure for lithium ion battery. *Solid State Ion.* **2018**, *315*, 102–110. [CrossRef]
32. Bashir, Z. Polyacrylonitrile, an unusual linear homopolymer with two glass transition. *Indian J. Fibre Text. Res.* **1999**, *24*, 1–9.
33. Wang, C.; Liu, Q.; Cao, Q.; Meng, Q.; Yang, L. Investigation on the structure and the conductivity of plasticized polymer electrolytes. *Solid State Ion.* **1992**, *53–56*, 1106–1110. [CrossRef]
34. Nicotera, I.; Coppola, L.; Oliviero, C.; Russo, A.; Ranieri, G.A. Some physicochemical properties of PAN-based electrolytes: Solution and gel microstructures. *Solid State Ion.* **2004**, *167*, 213–220. [CrossRef]
35. Enotiadis, A.; Fernandes, N.J.; Becerra, N.A.; Zammarano, M.; Giannelis, E.P. Nanocomposite electrolytes for lithium batteries with reduced flammability. *Electrochim. Acta* **2018**, *269*, 76–82. [CrossRef]
36. Ramesh, S.; Chiam, W.L. Rheological characterizations of ionic liquid-based gel polymer electrolytes and fumed silica-based composite polymer electrolytes. *Ceram. Int.* **2012**, *38*, 3411–3417. [CrossRef]
37. Rajendran, S.; Mahendran, O.; Kannan, R. Characterisation of [(1−x)PMMA–xPVdF] polymer blend electrolyte with Li+ ion. *Fuel* **2002**, *81*, 1077–1081. [CrossRef]
38. Sivakumar, M.; Subadevi, R.; Rajendran, S.; Wu, N.L.; Lee, J.Y. Electrochemical studies on [(1−x)PVA–xPMMA] solid polymer blend electrolytes complexed with LiBF$_4$. *Mater. Chem. Phys.* **2006**, *97*, 330–336. [CrossRef]

39. Cheng, S.H.S.; He, K.Q.; Liua, Y.; Zha, J.W.; Kamruzzaman, M.; Ma, R.L.W.; Dang, Z.M.; Li, R.K.Y.; Chung, C.Y. Electrochemical performance of all-solid-state lithium batteries using inorganic lithium garnets particulate reinforced PEO/LiClO$_4$ electrolyte. *Electrochim. Acta* **2017**, *253*, 430–438. [CrossRef]
40. Castriota, M.; Cazzanelli, E.; Nicotera, I.; Coppola, L.; Oliviero, C.; Ranieri, G.A. Temperature dependence of lithium ion solvation in ethylene carbonate–LiClO$_4$ solutions. *J. Chem. Phys.* **2003**, *118*, 5537. [CrossRef]
41. Zewde, B.W.; Carbone, L.; Greenbaum, S.; Hassoun, J. A novel polymer electrolyte membrane for application in solid state lithium metal battery. *Solid State Ion.* **2018**, *317*, 97–102. [CrossRef]
42. Zhang, H.; Liu, C.; Zheng, L.; Xu, F.; Feng, W.; Li, H.; Huang, X.; Armand, M.; Nie, J.; Zhou, Z. Lithium bis(fluorosulfonyl)imide/poly(ethylene oxide) polymer electrolyte. *Electrochim. Acta* **2014**, *133*, 529–538. [CrossRef]
43. Gorecki, W.; Jeannin, M.; Belorizky, E.; Roux, C.; Armand, M. Physical properties of solid polymer electrolyte PEO(LiTFSI) complexes. *Phys. Condens. Matter* **1995**, *7*, 34. [CrossRef]
44. Sumathipala, H.H.; Hassoun, J.; Panero, S.; Scrosati, B. High performance PEO-based polymer electrolytes and their application in rechargeable lithium polymer batteries. *Ionics* **2007**, *13*, 281–286. [CrossRef]

© 2018 by the authors. Licensee MDPI, Basel, Switzerland. This article is an open access article distributed under the terms and conditions of the Creative Commons Attribution (CC BY) license (http://creativecommons.org/licenses/by/4.0/).

Article

The Use of Succinonitrile as an Electrolyte Additive for Composite-Fiber Membranes in Lithium-Ion Batteries

Jahaziel Villarreal [1], Roberto Orrostieta Chavez [1], Sujay A. Chopade [2], Timothy P. Lodge [2] and Mataz Alcoutlabi [1,*]

[1] Department of Mechanical Engineering, University of Texas, Rio Grande Valley, Edinburg, TX 78539, USA; jahaziel.villarreal@lynntech.com (J.V.); roberto.orrostietachavez01@utrgv.edu (R.O.C.)
[2] Department of Chemical Engineering and Materials Science and Department of Chemistry, University of Minnesota, Minneapolis, MN 55455, USA; chopade123@gmail.com (S.A.C.); lodge@umn.edu (T.P.L.)
* Correspondence: mataz.alcoutlabi@utrgv.edu

Received: 13 January 2020; Accepted: 11 March 2020; Published: 17 March 2020

Abstract: In the present work, the effect of temperature and additives on the ionic conductivity of mixed organic/ionic liquid electrolytes (MOILEs) was investigated by conducting galvanostatic charge/discharge and ionic conductivity experiments. The mixed electrolyte is based on the ionic liquid (IL) (EMI/TFSI/LiTFSI) and organic solvents EC/DMC (1:1 v/v). The effect of electrolyte type on the electrochemical performance of a $LiCoO_2$ cathode and a SnO_2/C composite anode in lithium anode (or cathode) half-cells was also investigated. The results demonstrated that the addition of 5 wt.% succinonitrile (SN) resulted in enhanced ionic conductivity of a 60% EMI-TFSI 40% EC/DMC MOILE from ~14 mS·cm^{-1} to ~26 mS·cm^{-1} at room temperature. Additionally, at a temperature of 100 °C, an increase in ionic conductivity from ~38 to ~69 mS·cm^{-1} was observed for the MOILE with 5 wt% SN. The improvement in the ionic conductivity is attributed to the high polarity of SN and its ability to dissolve various types of salts such as LiTFSI. The galvanostatic charge/discharge results showed that the $LiCoO_2$ cathode with the MOILE (without SN) exhibited a 39% specific capacity loss at the 50th cycle while the $LiCoO_2$ cathode in the MOILE with 5 wt.% SN showed a decrease in specific capacity of only 14%. The addition of 5 wt.% SN to the MOILE with a SnO_2/C composite-fiber anode resulted in improved cycling performance and rate capability of the SnO_2/C composite-membrane anode in lithium anode half-cells. Based on the results reported in this work, a new avenue and promising outcome for the future use of MOILEs with SN in lithium-ion batteries (LIBs) can be opened.

Keywords: ionic liquids; succinonitrile; electrolyte; lithium ion batteries; composite fibers; mixtures

---

1. Introduction

Lithium-ion batteries (LIBs) are widely used in electronic devices ever since their successful commercialization by Sony in 1991 and Asahi Kasei and Toshiba in 1992 [1,2]. The conventional electrolyte used in LIBs is based on lithium hexafluorophosphate ($LiPF_6$) salt dissolved in volatile organic solvents; typically, these are mixtures of carbonates such as ethylene carbonate (EC), ethyl methyl carbonate (EMC), diethyl carbonate (DEC), and dimethyl carbonate (DMC) [3,4]. These combinations enhance desired properties in electrolytes. For example, EC has a high dielectric constant that promotes salt dissolution, and the addition of DMC to these organic solvents can lower the melting point and viscosity of the combined EC/DMC organic liquid electrolyte (OLE) [5]. Despite the high ionic conductivity and Li-ion diffusivity of OLEs during the charge/discharge cycles in LIBs, they face serious safety concerns due to their high flammability and volatility [6]. Such safety hazards can lead to thermal runaway and serious consequences [7,8]. Recently, incidents involving violent battery ignitions have caught the general public's attention, and this concern has increased even more with the application of lithium-ion batteries in electric vehicles and power grid storage devices [9]. Because ionic liquid electrolytes (ILEs) are non-flammable, non-volatile, and conductive, they possess safety advantages over OLEs and have been studied as electrolytes in rechargeable LIBs [10,11]. ILEs tend to be electrochemically and thermally stable, potentially allowing use of high voltage cathodes and safe operation at high temperatures. Nonetheless, ILEs face many challenges including larger viscosity, which increases significantly with decreasing temperature, crystallization at low temperatures, and large, strongly temperature-dependent interfacial impedance at both the cathode and anode. All of these issues are tied to the fact that the ILE-based Li-ion cells that have been developed to date operate only at elevated temperatures and at relatively low charging/discharging rates [10]. The high viscosity of ILEs typically results in a decreased total ionic conductivity at room temperature. Most important for electrolyte performance is to maximize the conductivity carried by the Li+ (cation), i.e., the product of total conductivity and transference number [12]. Additionally, it has been observed in both experiments and simulations that Li+ mobility increases more rapidly with dilution with an organic solvent than do the mobilities of the RTIL anion and especially cation, resulting in a higher transference number [13,14]. We note the further complication that the ability of a low viscosity solvent to dissolve lithium salt and transport Li^+ does not necessarily imply improved electrolyte performance [15,16].

A novel approach based on mixing organic solvents with ILEs has been used to improve the ionic conductivity of ILEs in Li-ion cells with the aim to offer combined advantages of OLs and ILEs such as decreased viscosity, higher Li+ diffusion/mobility, (i.e., improved conductivity), and, under appropriate volume ratios, better safety factors [6]. An increased tolerance to higher operational temperatures is beneficial not only because batteries are less likely to experience thermal runaway, but also because it enables specialty applications that require very low (−80 °C to −40 °C) or high (100 °C) temperatures; electrolyte crystallization can be prevented at lower working temperature conditions by modifying the IL solvent, while battery performance at higher temperature can be achieved by adding different lithium salts to ILs [17]. Several IL/Li salt systems have been employed as electrolytes in Li-ion batteries, but the number of ILEs that have been demonstrated as effective in operating cells is limited. Since demonstrating reasonable conductivity of an ILE (commonly done) and reasonable electrochemical stability (less commonly undertaken) are not sufficient to ensure reliable operation of a battery, a number of other issues such as interfacial defects must be considered. As a result, the number of operating batteries based on ILEs is less extensive than might be expected, given the huge number of anode/cathode and electrolyte combinations possible [11,18–20].

The most common anion investigated for ILEs is bis(trifluoromethanesulfonyl)-imide (TFSI), while the most common cations are alkylimidazoliums, tetraalkylammoniums, and alkylpryrrolidiniums (e.g., pyr13 and pyr14) [11]. 1-Ethyl-3-methylimidazolium bis(trifluoromethanesulfonyl)-imide (EMI-TFSI) has been used with Li salt as an electrolyte for LIBs due to its low viscosity compared to other ILEs [21]. For example, a $LiCoO_2$ cathode in a (EMI/TFSI + LiTFSI) electrolyte delivered higher discharge capacity at room temperature than in the $EMIBF_4$ + $LiBF_4$ system [21]. EMI-TSFI is employed here as an IL due to its good conductivity (8.7–9.1 mS·cm^{-1} at 25 °C), low viscosity (33–34 cP), and low melting point (−15 °C) [17]. Additionally, it has been found that EMI-TSFI with LiTSFI can increase the ionic conductivity and decrease the viscosity of the electrolyte, while no flammability was observed for compositions with IL (EMI-TSFI) wt% of 40% or more in EC–DEC–VC–1M $LiPF_6$ electrolytes at increased temperatures [21,22].

Reversible capacities of up to 155 mAh g^{-1} and 128 mAh g^{-1} have been reported for Li-ion full cells with EMI-TSFI and EC/DMC MOILEs using $LiFePO_4$ as the cathode with a graphite anode, and a $LiFePO_4$ cathode with a $Li_4Ti_5O_{12}$ anode, respectively [22]. Nonetheless, the addition of non-ionic organic additives such as succinonitrile (SN) to the electrolyte can improve the ionic conductivity of ILEs and the overall electrochemical performance of the Li-ion cell [1,23]. SN can dissolve different types of salts such as LiTFSI, $LiBF_4$, $LiPF_6$, $LiN(CN)_2$, $Ba(TFSI)_2$, $Pb(TSFI)_2$, $La(TSFI)_2$, $Ag(CF_3SO_3)$, and $Cu(CF_3SO_3)$ [24]. SN has been frequently used as a solid electrolyte in LIBs [24–28] but limited results have been reported on the use of SN with OLEs and ILEs in LIBs [29]. For example, the addition of SN to a polymer electrolyte (PEO-LiTFSI, P(VDFHFP)– LiTFSI, and P(VDF-HFP)–LiBETI) resulted in improved ionic conductivity of the polymer electrolyte and favorable mechanical properties [30]. SN has been recently used as a functional additive to improve the thermal stability and broaden the oxidation electrochemical window of an OLE in lithium cathode half-cells containing a LNMO cathode. The results showed that the addition of SN to the electrolyte solution lead to a remarkably improved cycling stability, which was due to the formation of an electronically conductive film on the cathode [29,30]. SN was also used as an additive to improve the thermal stability of ethylene carbonate (EC)-based electrolytes in LIBs. This work showed that SN can suppress parasitic reactions between the positive electrode (LiCoO$_2$) and the organic liquid electrolyte, because the nitrogen ion in the nitrile functional group (–CN) in SN has a lone pair of electrons leading to a strong bond with the transition metal ions on the cathode. It was also reported that the addition of SN resulted in a suppression of electrolyte decomposition in commercial cells [30]. The authors also suggested that SN can react with transition metal ions in the electrolyte to form metal ion compounds preventing their reduction on the negative electrode surface, which would compromise the SEI surface. This work focuses on the investigation of the electrochemical properties of EMI-TFSI-LiTFSI electrolytes in lithium anode (or cathode) half-cells using either a cathode or an anode [31]. OLE (EC/DMC 1:1 v/v), ILE (EMI-TFSI), and MOILE were used with a commercial cathode, LiCoO$_2$, and a SnO$_2$/C composite-fiber anode in lithium anode half-cells to investigate the effect of electrolyte type on the electrochemical performance. The effects of temperature and SN additive on the ionic conductivity and electrochemical performance of MOILEs were investigated by conducting charge/discharge and impedance measurements on the lithium anode (or cathode) half-cells with commercial cathode materials.

2. Experimental

2.1. Materials

Bis(trifluoromethane)sulfonimide lithium salt (Li-TFSI) (99%) and 1-ethyl-3-methylimidazolium bromide (EMI-Br) (99.98%) were purchased from Io-li-tech, Tuscaloosa, AL, USA. Ethylene carbonate (EC) (99%) and dimethyl carbonate (DMC) were purchased from Alpha Aesar (Tewksbury, MA, USA) and Fisher Scientific (Lenexa, KS, USA), respectively. LiTFSI (98%), HPLC water, silicon oxide, dichloromethane (DCM) (99%), and succinonitrile (SN, 99%) were obtained from Sigma-Aldrich (St. Louis, MO, USA)). Polyacrylonitrile (PAN) with $M_w \approx 150{,}000$ was purchased from Sigma-Aldrich. Dimethylformamide (DMF) (>99.5%), and tin (II) 2-ethylhexaonate were purchased from Fisher Scientific.

2.2. Electrolyte Preparation

1-Ethyl-3-methyl-imidazolium bis(trifluoromethanesulfonyl)imide (EMI-TFSI), was synthesized by reacting HPLC water with a 1:1 LiTFSI:EMI-Br molar ratio mixture. This solution was stirred for 24 h in an oil bath at 70 °C. After the reaction took place, an aqueous layer and an ionic liquid (EMI-TFSI) rich layer were formed, and the solution was extracted from the oil bath. Once the solution was cooled to room temperature, the EMI-TFSI was separated from its aqueous counterpart and decanted into a separator funnel. HPLC water was poured into the separator funnel and mixed with the EMI-TFSI. The mixture was left to rest until the two layers were formed again. The EMI-TFSI layer was once again removed. This process was repeated two more times. Then, the EMI-TFSI was placed in a 500 mL round bottom flask to be dissolved with a sufficient amount of DCM. The dissolved EMI-TSFI was decanted into a chromatography column in order to filter any remaining impurities. The chromatography column contained one inch of sand, followed by silica oxide fully covering the remaining of the column up to the beginning of reservoir. The collected solution was then placed in a rotavap to remove the solvent (DCM) from the EMI-TSFI. Finally, the obtained EMI-TSFI was placed in a vacuum oven at 100 °C for 48 h to remove any water and excess DCM. The purity of the synthesized IL electrolyte was confirmed by ^{1}H NMR spectroscopy.

The organic liquid electrolyte was prepared in a glove box (MBRAUN, Garching, Germany) with a controlled argon atmosphere. A 20 mL solution was prepared by combining a 1:1 v/v ratio of EC and DMC followed by 2 h of magnetic stirring. This OLE solution was stored and used to make 5 mL batches of 1 M LiTFSI in 60% EMI-TFSI and 40% EC/DMC. First, 1.435 g of LiTFSI and 2.564 g of EC/DMC solution were stirred until the LiTFSI was fully dissolved. Next, 4.590 g of EMI-TFSI were added and stirred for 24 h. The final weight of the solution was 8.590 g. Using this weight, an additional 0.429 g of SN was added to compare the ionic conductivity of the MOILE with one containing 5 wt.% of SN.

The ionic conductivity of the MOILEs was measured by assembling coin-type cells (CR2032) composed of two stainless-steel spacers as the positive and negative terminals, and a Teflon washer filled with MOILE. The LiCoO$_2$ cathode was assembled with a common half-cell configuration to investigate the electrochemical performance of the cell. The as-prepared MOILE was used with the commercial LiCoO$_2$ cathode. The active material loading in the electrode was 6.2–8.0 mg/cm^2. The coin cells were assembled in a glove box using the cathode as the working electrode, with a Li counter electrode and microfiber glass mat separator (Whatman).

2.3. Preparation of SnO$_2$/C Composite Fiber Membranes

The SnO$_2$/C composite fibers were prepared by forcespinning of PAN/SnO$_2$ precursor fibers followed by a thermal treatment. The PAN/SnO$_2$ solution was prepared by dissolving 12 wt% PAN in DMF. A tin (II) 2-ethylhexanoate solution to 2:1 weight ratio of PAN solution was added and stirred for 24 h. The forcespinning process relies on applying centrifugal forces at high rotational speeds to a polymer solution or melt to produce microfibers with different structure and morphology. A description of the forcespinning process was given previously [19,32,33]. The PAN/SnO$_2$ precursor solution was spun using the FiberRio L-1000 cyclone at a rotational speed of 8000 rpm for 1 min. The PAN/SnO$_2$ fibrous mat was collected, stabilized in air at 280 °C for 5 h, and subsequently carbonized under argon atmosphere at 700 °C for 3 h (heating rate: 3 °C/min). The SnO$_2$/C composite fibers were removed from the tube furnace, punched into 0.5 in (0.0127 m) diameter anodes, then weighed and used directly as working electrodes in lithium anode half-cells.

2.4. Fiber Membrane Characterization

The morphology and structure of composite fiber membranes were investigated using a scanning electron microscope (SEM) from Sigma VP Carl Zeiss, Oberkochen, Germany while energy-dispersive X-ray spectroscopy (EDS) from EDAX Inc., Mahwah, NJ, USA was used to confirm the elemental composition of the fibers. The crystal structure of the composite fiber membranes was evaluated by X-ray diffraction (XRD) using a Bruker D8 Advanced X-ray Diffractometer at a scan rate of 1 °C/min over a range of 2θ angle from 10° to 70°.

2.5. Electrochemical Measurements

Lithium anode (or cathode) half-cells were assembled in an argon-filled glove box with SnO$_2$/C composite fibers as a binder-free anode and Li-metal as the counter electrode, using the MOILE. The electrothermal performance of the SnO$_2$/C composite-fiber anode was evaluated by conducting galvanostatic charge/discharge experiments on CR2032 coin cells at 100 mA g^{-1}. The active material loading in the anode was 2.4–4.5 mg/cm^2. The ionic conductivity experiments on half cells with MOILEs were performed at different temperatures using a home-built heating block chamber. The design was based on a home-built sealed conducting cell in use at the University of Minnesota [23]. The impedance of the MOILEs at different temperatures was measured using a Metrohm Autolab (PGSTAT128N) connected to the heating chamber, over a frequency range from 0.1 Hz to 1 kHz. The ionic conductivity of the electrolyte was determined using coin cells with two stainless steel blocking electrodes filled with the electrolyte. For accurate measurements of the ionic conductivity, a Teflon spacer was placed between the stainless-steel electrodes to hold the electrolyte inside the cell. The sample (electrolyte) preparation was conducted in an argon-filled glove box. The cell was then taken outside the glove box and inserted in the heating chamber. The ionic conductivity, σ, was calculated as L/(RA), where L and A are the sample thickness and superficial area of the sample and R is the bulk resistance [23]. The bulk resistance was determined from the frequency-independent plateau of the real part of the impedance (Z′). The temperature was controlled and monitored using thermocouples and heating cartridges connected to a temperature process control CN 7500 purchased from Omega. The experimental setup was connected to a personnel computer using a RS485 USB converter to monitor the time and temperature during the impedance measurements.

The electrochemical performance of the $LiCoO_2$ half-cells was evaluated at 60 °C. The $LiCoO_2$ half cells were placed in a controlled temperature oven (ESPEC BTZ – 133). $LiCoO_2$ cathode with electrolytes 1 M LiTFSI in 60% EMI-TFSI and 40% EC/DMC with and without SN were tested at a current density of 100 mAh g^{-1} for 50 cycles. Arbin's MTIS Pro was employed to conduct the galvanostatic charge/discharge experiments over a voltage range of 2.5–4.2 V. A port extension was connected between the Arbin instrument and the ESPEC oven to conduct the electrochemical experiments at different temperatures.

3. Results and Discussion

3.1. Materials Characterization

Figure 1 shows SEM images of SnO_2/C composite fibers. It can be seen in Figure 1 that the SnO_2 nanoparticles tend to aggregate, forming large clusters on the fibers. Some of these nanoparticles are embedded in the fibers and some are deposited on the fiber strands [20]. The average fiber diameter of the SnO_2/C composite fibers was 1.86 m.

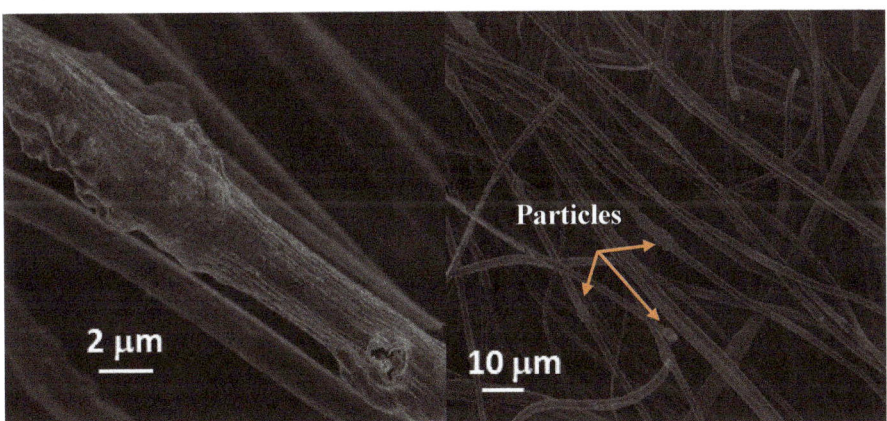

Figure 1. SEM images of a SnO_2/C composite-fiber membrane [20], with copyright permission from the IOP Publishing.

Figure 2 shows an SEM image of SnO_2/C composite fibers and the corresponding EDS mapping. Figure 2 shows that the composite fibers consist of C, O, and Sn that are distributed over the fibers. The EDS results confirm that the aggregated nanoparticles on the fibers contain Sn and O (i.e., SnO_2 nanoparticles), which are attached to the surface of the carbon-fiber matrix

Figure 2. SEM image of SnO$_2$/C composite fibers (left) and corresponding EDS mapping of the SnO$_2$/C composite fibers (right).

Figure 3 shows an XRD pattern for the carbon fibers, where a broad diffraction peak is observed at 2θ = 27.8° corresponding to the (002) lattice plane of graphite [34–36]. It is observed in Figure 3 that this peak is weak and broad, which is the result of the formation of an amorphous carbon fiber structure after carbonization of the precursor PAN fibrous membrane.

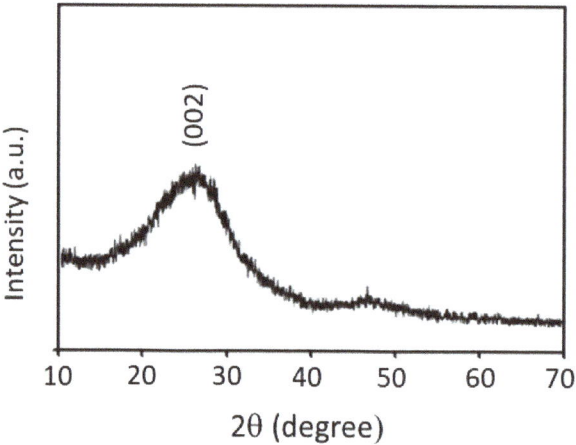

Figure 3. XRD pattern of the carbon-fiber membrane prepared after the carbonization of polyacrylonitrile (PAN) fibers at 700 °C.

Figure 4 shows XRD analysis of the SnO$_2$/C composite-fiber membrane. The observed pattern has predominantly crystalline peaks corresponding to (110), (101), (200), (211), and (310) planes. The observed peaks overlap with five of the seven peaks of the SnO$_2$ crystal structure published by the (JCPDS 41-1445), further confirming the formation of SnO$_2$ nanoparticles in the carbon matrix.

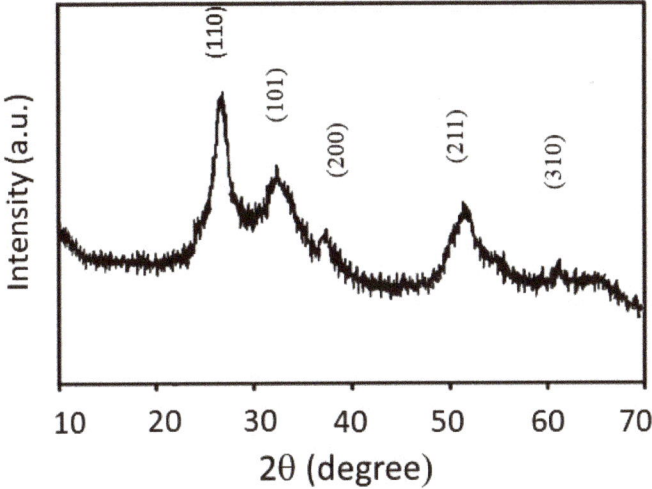

Figure 4. XRD pattern of the SnO$_2$/C composite-fiber membrane prepared after calcination of PAN/SnO$_2$ precursor fibers at 700 °C.

3.2. Ionic Conductivity Measurement of Electrolytes at Different Temperature

Figure 5 shows the ionic conductivity of the ILE and MOILEs as a function of temperature. The ILE was prepared from LiTFSI salt dissolved in 100% EMI-TFSI while the MOILE was prepared by dissolving LiTFSI salt in 60% EMI-TFSI and 40% EC/DMC, with and without the addition of 5% SN. The results show that the ILE (100% EMI-TFSI) delivered an ionic conductivity ~5 mS·cm^{-1} at room temperature, which is lower than that for the MOILE (60% EMI-TFSI and 40% EC/DMC) (~14 mS·cm^{-1}). It is also clear in Figure 5 that the ionic conductivity of the three electrolytes increases with increasing temperature. The MOILE with 5% SN shows the highest ionic conductivity at 100 °C (70 mS·cm^{-1}) among these three electrolytes. Despite its lower conductivity at room temperature, the ILE ionic conductivity increased significantly as the temperature was increased. At 150 °C, the ILE conductivity was ~30 mS·cm^{-1}. This behavior is expected since the viscosity of ILEs tends to decrease with increasing temperature. The addition of 40% organic liquid, EC/DMC (1:1 v/v ratio), to 1 M LiTFSI in 60% EMI-TFSI resulted in an increased ionic conductivity of ~14 mS·cm^{-1} at room temperature while the MOILE with the addition of 5 wt% SN exhibited the highest room temperature ionic conductivity of ~26 mS·cm^{-1}. Note here that the ionic conductivity of the OLE (EC/DMC/LiTFSI) is not shown in Figure 5 since there are data available in the literature on LiTFSI in binary EC/DMC mixtures. In fact, LiTFSI salt in EC/DMC binary system shows a higher ionic conductivity than that for ILE and MOILE. For example, results reported by Dahbi et al. showed that the LiTFSI in EC/DMC (1:1 v/v ratio), which is the same OLE used in the present work, exhibited an ionic conductivity of 8.6 mS·cm^{-1} at 25 °C. This value was increased to 11.5 and 14.9 mS·cm^{-1} when the temperature was increased to 40 and 60 °C, respectively [37]. The results also showed the ionic conductivity of LiPF$_6$ in EC/DMC mixture was higher than that with EC/DMC/LiTFSI electrolyte over the entire temperature range [37].

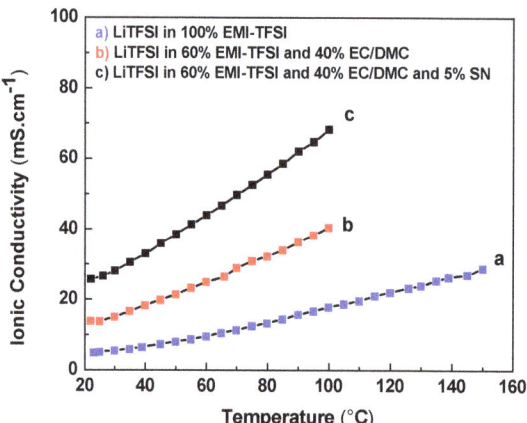

Figure 5. Ionic conductivity vs temperature for the ionic liquid electrolyte (ILE), mixed organic/ionic liquid electrolyte (MOILE), and MOILE with 5 wt.% SN.

3.3. Electrochemical Performance of A LiCoO$_2$ Electrode in Lithium Cathode Half-Cells

The commercial LiCoO$_2$ electrode was employed in lithium cathode half-cells with a single-coated lithium foil to investigate its electrochemical performance. The MOILE with and without 5 wt% SN was used with the commercial LiCoO$_2$. The purpose was to evaluate the behavior of the MOILEs in high voltage cathode materials such as LiCoO$_2$, which has a larger voltage range than LiFePO$_4$. LiCoO$_2$ still dominates the portable electronics market due to its high voltage plateau and easy synthesis compared to LiFePO$_4$ [38]. Galvanostatic charge/discharge experiments were performed for 50 cycles at different temperatures and at a current density of 100 mA g^{-1}.

Figure 6a,b shows the charge/discharge profiles at 60 °C and at 100 mA g^{-1} of the commercial LiCoO$_2$ cathode in MOILEs without SN and with 5 wt% SN, respectively. As can be observed in Figure 6a, the LiCoO$_2$ cathode in the MOILE without SN maintained a consistent specific capacity of 148 mAh g^{-1} up to the 10th cycle. However, significant irreversibilities were observed at the 25th and 50th cycles. After 50 cycles, the cathode delivered a discharge capacity of 91 mAh g^{-1}, indicating a capacity retention of 61.5% at a current density of 100 mA g^{-1}. The discharge capacity retention is equal to the capacity after the 50th cycle divided by the capacity at the 1st cycle (i.e., 61.5% = (100−38.5)%). On the other hand, the LiCoO$_2$ cathode in the MOILE with SN (Figure 6b) exhibited an initial discharge capacity of 150 mAh g^{-1} at 100 mA g^{-1}, and after the 50th cycle, the discharge capacity reached a value of 129 mA g^{-1} indicative of acceptable capacity retention of 86%. The improvement in the electrochemical performance of the LiCoO$_2$ cathode is attributed to the effect of the SN additive on the ILE, and to the high conductivity of MOILEs at high temperature (60 °C). The high volatility and evaporation (high vapor pressure) of DMC at high temperature might influence the ionic conductivity of electrolytes containing a high percentage of DMC, thus affecting the electrochemical performance of the electrode. The effect of DMC on the ionic conductivity of MOILEs was not investigated since the amount of DMC in MOILEs is only 20% (1 M LITFSI in 1:1 v/v EC/DMC) and this should affect the performance of the electrode only slightly. However, results reported in the literature show that the ionic conductivity of 1 M LiPF$_6$ in DMC remains significant (i.e., 9 mS·cm^{-1}) at 55 °C [39]. Results reported by Aurbach et al. on a LNMO cathode at 60 °C in a liquid electrolyte (DMC–EC (2:1)/LiPF$_6$ 1.5 M), over a 3.5–4.9 V potential range showed that the cycling behavior of the cathode was explored without any observed degradation of the electrolyte solution [40].

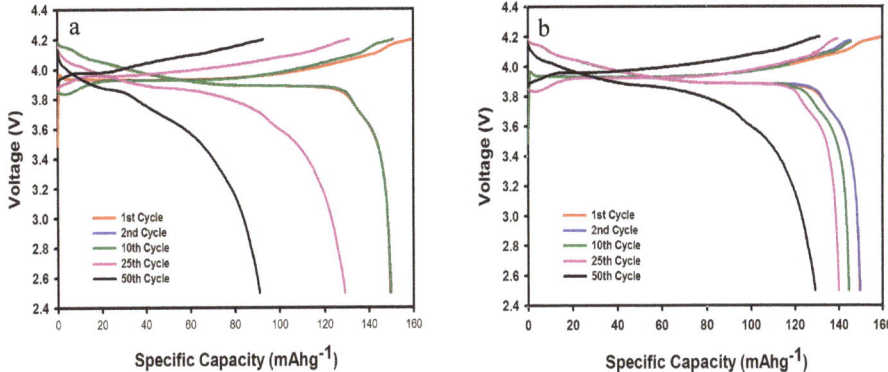

Figure 6. Charge/discharge profiles of a commercial LiCoO$_2$ cathode at 60 °C with MOILEs (**a**) 1 M LiTFSI 60% EMI-TFSI 40% EC/DMC EC/DMC (1:1 v/v), and (**b**) 1 M LiTFSI 60% EMI-TFSI 40% EC/DMC EC/DMC (1:1 v/v) containing 5 wt% SN. Current density = 100 mA g^{-1}.

It is worth noting here that the LiCoO$_2$ cathode in the MOILEs shows moderate capacity fading and voltage change in the plateau of Figure 6a,b. This might be caused by a decrease in active material (lithium) on the current collector after the 25th cycle. Another important factor that could affect this loss in capacity of the LiCoO$_2$ cathode is that the corrosion of the Al current collector on the cathode side by the TFSI, thereby contributing to the loss of active material from the Al current collector [41]. More work will be conducted to investigate these effects on the LiCoO$_2$ cathode in LiTFSI/MOILEs systems.

Figure 7a,b shows the cycling performance corresponding to the charge/discharge curves shown in Figure 6a,b. Although the capacity is stable within the first twenty cycles, the LiCoO$_2$ cathode in the MOILE without SN suffered from a steady loss in specific capacity after 20 cycles. In contrast, the same cathode in the MOILE with 5 wt% SN maintained a stable specific capacity for the first 20 cycles; there was a slight decrease in capacity between 20th and 30th cycles, while thereafter the cathode maintained a constant capacity of ~129 mAh g^{-1}. The LiCoO$_2$ cathode in both electrolytes maintained a similar high coulombic efficiency of 98% for 50 cycles.

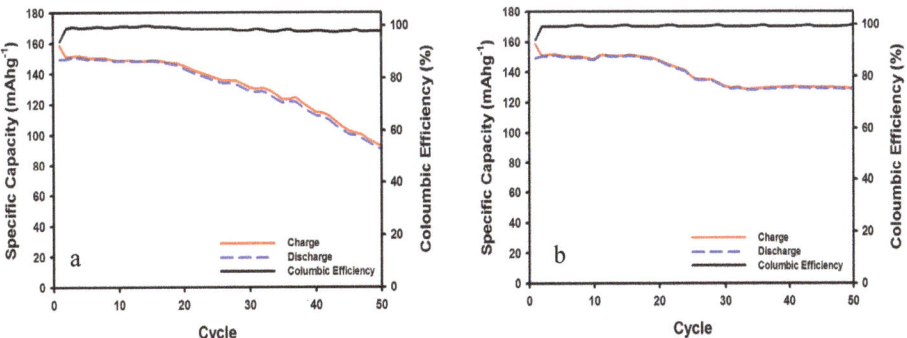

Figure 7. Cycling performance and coulombic efficiency of LiCoO$_2$ commercial cathode at 60 °C in (**a**) 1 M LiTFSI in 60% EMI-TFSI 40% EC/DMC (1:1 v/v) electrolyte, and (**b**) 1 M LiTFSI in 60% EMI-TFSI 40% EC/DMC (1:1 v/v) with 5 wt% SN. Current density = 100 mA g^{-1}.

Figure 8a,b shows the charge/discharge curves at 100 mA g^{-1} for the SnO$_2$/C composite-fiber anode in two different electrolytes, OLE and MOILE with SN. The cycle performance of the SnO$_2$/C composite electrode was evaluated by conducing galvanostatic charge/discharge experiments at room temperature and at a current density of 100 mA g^{-1}. The voltage range for lithium anode half-cells tested with the SnO$_2$/C composite-fiber anode was 0.05–3 V (versus Li+/Li). The SnO$_2$/C composite-fiber anode with the 1 M LiPF$_6$ in EC/DMC (1:1 v/v) electrolyte showed an initial discharge capacity of 785 mAh g^{-1}. The reversible specific capacity after 100 cycles was 319 mAh g^{-1}. Nevertheless, the SnO$_2$/C composite anode showed a stable specific capacity after the 25th cycle, with a capacity retention of 98% after the 2nd cycle. Improved cycling stability of the SnO$_2$/C composite fibers in 1 M LiTFSI in 60% EMI-TFSI 40% EC/DMC with 5% SN electrolyte (Figure 8b) was observed after the 2nd cycle, with a specific capacity of 382 mAh g^{-1}, having a ~20% increase compared to the SnO$_2$/C composite-fiber anode cycled with the OLE.

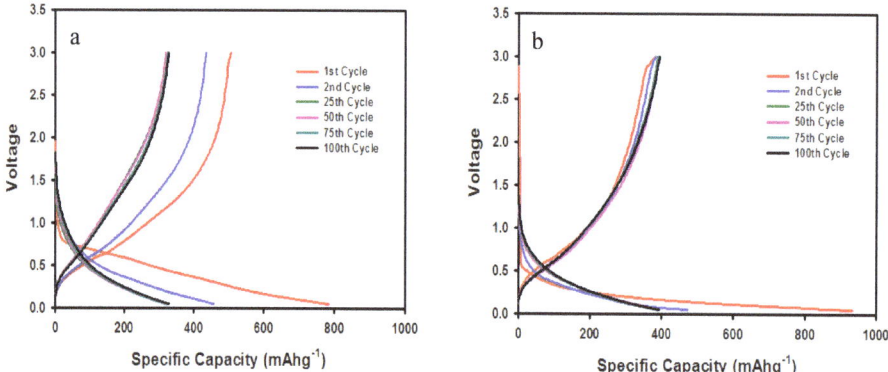

Figure 8. Charge/discharge profiles for a SnO$_2$/C composite-fiber anode at 25 °C in two different electrolytes: (**a**) 1M LiPF$_6$ in EC/DMC (1:1 v/v) electrolyte, and (**b**) 1 M LiTFSI in 60% EMI-TFSI/ 40% EC/DMC (1:1 v/v) with 5 wt% SN.

Figure 9 shows the cycling performance (charge/discharge capacity vs. cycle number) of the SnO$_2$/C composite-membrane anode in the OLE and the MOILE with 5 wt% SN at a current density of 100 mA g^{-1}. It is observed in Figure 9a,b that the discharge and charge capacities of the SnO$_2$/C composite-fiber anode in the MOILE with 5 wt% SN are higher than in the OLE. The improvement in the specific capacity of the composite-membrane anode was attributed to the addition of the high polarity SN to the MOILE and its ability to dissolve the LiTFSI salt, which resulted in enhanced ionic conductivity and improved cycling stability of the electrode in the MOILE. The SnO$_2$/C composite-membrane anode in MOILE with 5 wt.% SN shows (Figure 9b) improved cycling stability and capacity retention after the 2nd cycle.

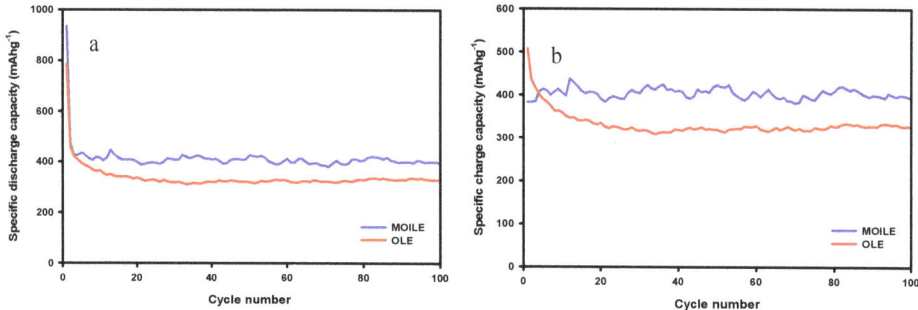

Figure 9. Cycling performance of a SnO$_2$/C composite-fiber anode at 25 °C in two different electrolytes: (**a**) 1M LiPF$_6$ in EC/DMC (1:1 v/v) electrolyte, and (**b**) 1 M LiTFSI in 60% EMI-TFSI/ 40% EC/DMC (1:1 v/v) with 5 wt% SN.

The rate performance of the SnO$_2$/C composite fibers was further evaluated by conducting current rate (or rate capability) experiments on the lithium anode half-cells at different current densities between 50 and 500 mA g^{-1}. The SnO$_2$/C composite fibers were cycled ten times at current densities of 50, 100, 200, 400, 500, and then again at the initial value of 50 mA g^{-1} (Figure 10). The results exemplify the SnO$_2$/C composite anode's ability to perform at higher current densities, as well as demonstrating the capacity recovery after being cycled from high to low current density. Figure 10 shows the rate performance (charge capacity vs cycle number at different current densities) of the SnO$_2$/C composite-fiber anode in the OLE and in the MOILE with 5 wt.% SN. As expected, the composite-fiber anode delivered a higher specific charge capacity at lower current density, and vice versa. At 50 mAh g^{-1}, the specific capacity decreased after 10 cycles to 418 mAh g^{-1} for the Li-ion cell cycled with the 1 M LiPF$_6$ in EC/DMC (1:1 v/v) electrolyte, but only to 579 mAh g^{-1} for the 1 M LiTFSI in 60% EMI-TFSI 40% EC/DMC (1:1 v/v) with 5 wt% SN electrolytes. This can be attributed to the stresses and strains caused by the high-volume change of the SnO$_2$/C composite fibers after repeated charge/discharge cycles. At a current density of 100 mA g^{-1}, the charge capacity was stable at ~315 mAh g^{-1} for 1 M LiPF$_6$ in EC/DMC (1:1 v/v) and at ~441 mAh g^{-1} for 1 M LiTFSI in 60% EMI-TFSI 40% EC/DMC (1:1 v/v) 5 wt% SN. The SnO$_2$/C composite anode in the MOILE with 5 wt% SN had a higher percentage increase in specific capacity with 25% at 100 mA g^{-1}, 23% at 200 mA g^{-1}, 30% at 400 mA g^{-1}, and 1% at 500 mA g^{-1}, compared to 1 M LiPF$_6$ in EC/DMC (1:1 v/v). However, after cycling back to 50 mA g^{-1}, the SnO$_2$/C composite fibers with 1 M LiPF$_6$ in EC/DMC (1:1 v/v) had less specific charge capacity than with the MOILE with 5 wt% SN. However, the SnO$_2$ electrode in both electrolytes (OLE and MOILE with SN) shows relatively low capacity at higher current density. Thus, the improvement in the charge capacity of the SnO$_2$/C composite anode with MOILE and SN can be attributed to the high Li-ion conductivity and diffusion caused by the addition of SN to the ionic liquid electrolyte.

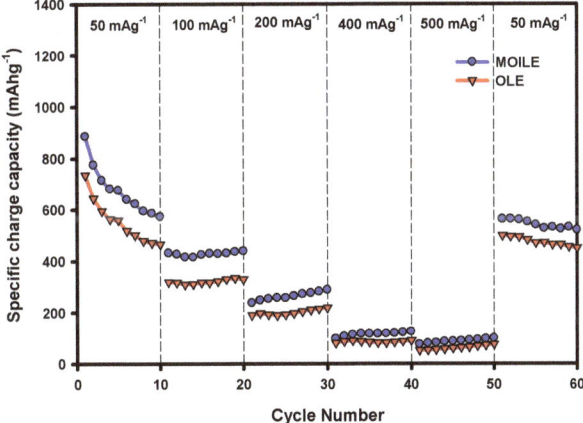

Figure 10. Rate performance (charge capacity vs cycle number at different current densities) of SnO$_2$/C composite fibers at 25 °C with two different electrolytes: OLE (1M LiPF$_6$ in EC/DMC (1:1 v/v)), and MOILE (1 M LiTFSI in 60% EMI-TFSI/ 40% EC/DMC (1:1 v/v) with 5 wt% SN).

4. Conclusions

Two different electrolytes, (1 M LiPF$_6$ in EC/DMC (1:1 v/v) and 1 M LiTFSI in 60% EMI-TFSI 40% EC/DMC (1:1 v/v) with 5 wt% SN), were synthesized, characterized electrochemically, and compared using lithium anode (or cathode) half-cells with either a SnO$_2$/C anode or a LiCoO$_2$ cathode. The SnO$_2$/C composite-fiber electrode was prepared by forcespinning of a PAN/SnO$_2$ precursor solution and subsequent thermal treatment. The electrochemical performance results showed that lithium anode half-cells with a SnO_2/C composite-fiber electrode in 1 M LiTFSI in 60% EMI-TFSI 40% EC/DMC (1:1 *v/v*) and 5 wt% SN perform better than that with commercial organic liquid electrolyte. The use of ionic liquid electrolyte with 5 wt% SN in lithium anode half-cells with a SnO_2/C electrode demonstrated good cycling stability and capacity retention after 100 charge/discharge cycles. The results showed that 1 M LiTFSI in 60% EMI-TFSI 40% EC/DMC (1:1 *v/v*) with 5 wt% SN had a higher ionic conductivity than 1 M $LiPF_6$ in EC/DMC (1:1 *v/v*) electrolyte. The electrochemical performance of a commercial $LiCoO_2$ cathode was evaluated at 60 °C using lithium cathode half-cells with MOILEs, both without and with SN electrolytes. The commercial $LiCoO_2$ cathode was evaluated electrochemically at 60 °C, cycled with 1 M LiTFSI in 60% EMI-TFSI 40% EC/DMC (1:1 *v/v*) and 5 wt% SN and had an excellent performance. The $LiCoO_2$ cathode in 60% EMI-TFSI 40% EC/DMC (1:1 *v/v*) and 5 wt% SN showed good electrochemical performance at 60 °C, which was attributed to the high ionic conductivity of the MOILE/SN at elevated temperature.

Author Contributions: For research articles with several authors, a short paragraph specifying their individual contributions must be provided. The following statements should be used "Conceptualization, M.A.methodology, S.A.C. and J.V.; validation, J.V. and and S.A.C.; formal analysis, J.V.; investigation, J.V. and S.A.C.; data curation and analysis, J.V. and R.O.C.; writing J.V.; R.O.C. and M.A.;—original draft preparation, J.V., and M.A.; writing—review and editing, M.A. and T.P.L.; supervision, M.A. and T.P.L.; funding acquisition, M.A. and T.P.L. All authors have read and agreed to the published version of the manuscript."

Funding: This research was supported by NSF PREM award under grant No. DMR-1523577: UTRGV-UMN Partnership for Fostering Innovation by Bridging Excellence in Research and Student Success. Part of this work was carried out in the College of Science and Engineering Characterization Facility, University of Minnesota, which has received capital equipment funding from the NSF through the UMN MRSEC program under Award Number DMR-1420013.

Acknowledgments: This research was supported by NSF PREM award under grant No. DMR-1523577: UTRGV-UMN Partnership for Fostering Innovation by Bridging Excellence in Research and Student Success. Part of this work was carried out in the College of Science and Engineering Characterization Facility, University of Minnesota, which has received capital equipment funding from the NSF through the UMN MRSEC program under Award Number DMR-1420013.

Conflicts of Interest: The authors declare no conflict of interest.

References

1. Tarascon, J.M.; Armand, M. Issues and challenges facing rechargeable lithium batteries. *Nature* **2001**, *414*, 359–367. [CrossRef] [PubMed]
2. Goodenough, J.B.; Park, K.S. The Li-Ion Rechargeable Battery: A Perspective. *J. Am. Chem. Soc.* **2013**, *135*, 1167–1176. [CrossRef] [PubMed]
3. Xu, K. Nonaqueous liquid electrolytes for lithium-based rechargeable batteries. *Chem. Rev.* **2004**, *104*, 4303–4417. [CrossRef]
4. Marcinek, M.; Syzdek, J.; Marczewski, M.; Piszcz, M.; Niedzicki, L.; Kalita, M.; Plewa-Marczewska, A.; Bitner, A.; Wieczorek, P.; Trzeciak, T.; et al. Electrolytes for Li-ion transport—Review. *Solid State Ionics* **2015**, *276*, 107–126. [CrossRef]
5. Kalhoff, J.; Eshetu, G.G.; Bresser, D.; Passerini, S. Safer Electrolytes for Lithium-Ion Batteries: State of the Art and Perspectives. *Chemsuschem* **2015**, *8*, 2154–2175. [CrossRef]
6. Quinzeni, I.; Ferrari, S.; Quartarone, E.; Tomasi, C.; Fagnoni, M.; Mustarelli, P. Li-doped mixtures of alkoxy-N-methylpyrrolidinium bis(trifluoromethanesulfonyl)-imide and organic carbonates as safe liquid electrolytes for lithium batteries. *J. Power Sources* **2013**, *237*, 204–209. [CrossRef]
7. Feng, X.N.; Zheng, S.Q.; Ren, D.S.; He, X.M.; Wang, L.; Liu, X.; Li, M.G.; Ouyang, M.G. Key Characteristics for Thermal Runaway of Li-ion Batteries. *Energy Proced.* **2019**, *158*, 4684–4689. [CrossRef]
8. Balakrishnan, P.G.; Ramesh, R.; Kumar, T.P. Safety mechanisms in lithium-ion batteries. *J. Power Sources* **2006**, *155*, 401–414. [CrossRef]
9. Feng, X.N.; Ouyang, M.G.; Liu, X.; Lu, L.G.; Xia, Y.; He, X.M. Thermal runaway mechanism of lithium ion battery for electric vehicles: A review. *Energy Storage Mater.* **2018**, *10*, 246–267. [CrossRef]
10. Navarra, M.A. Ionic liquids as safe electrolyte components for Li-metal and Li-ion batteries. *MRS Bull.* **2013**, *38*, 548–553. [CrossRef]

11. Lewandowski, A.; Swiderska-Mocek, A. Ionic liquids as electrolytes for Li-ion batteries-An overview of electrochemical studies. *J. Power Sources* **2009**, *194*, 601–609. [CrossRef]
12. Molinari, N.; Mailoa, J.P.; Kozinsky, B. General Trend of a Negative Li Effective Charge in Ionic Liquid Electrolytes. *J. Phys. Chem. Lett.* **2019**, *10*, 2313–2319. [CrossRef]
13. Li, Z.; Smith, G.D.; Bedrov, D. Li+ Salvation and Transport Properties in Ionic Liquid/Lithium Salt Mixtures: A Molecular Dynamics Simulation Study. *J. Phys. Chem. B* **2012**, *116*, 12801–12809. [CrossRef]
14. Bayley, P.M.; Lane, G.H.; Rocher, N.M.; Clare, B.R.; Best, A.S.; MacFarlane, D.R.; Forsyth, M. Transport properties of ionic liquid electrolytes with organic diluents. *Phys. Chem. Chem. Phys.* **2009**, *11*, 7202–7208. [CrossRef]
15. Borodin, O.; Henderson, W.A.; Fox, E.T.; Berman, M.; Gobet, M.; Greenbaum, S. Influence of Solvent on Ion Aggregation and Transport in PY15TFSI Ionic Liquid-Aprotic Solvent Mixtures. *J. Phys. Chem. B* **2013**, *117*, 10581–10588. [CrossRef]
16. Dong, D.P.; Bedrov, D. Charge Transport in [Li(tetraglyme)][bis(trifluoromethane) sulfonimide] Solvate Ionic Liquids: Insight from Molecular Dynamics Simulations. *J. Phys. Chem. B* **2018**, *122*, 9994–10004. [CrossRef]
17. Wasserscheid, P.; Welton, T. *Ionic Liquids in Synthesis*; Wiley-VCH: Hoboken, NJ, USA, 25 June 2008; Volume 1.
18. Ji, L.W.; Lin, Z.; Alcoutlabi, M.; Zhang, X.W. Recent developments in nanostructured anode materials for rechargeable lithium-ion batteries. *Energy Environ. Sci.* **2011**, *4*, 2682–2699. [CrossRef]
19. Agubra, V.A.; Zuniga, L.; Flores, D.; Villareal, J.; Alcoutlabi, M. Composite Nanofibers as Advanced Materials for Li-ion, Li-O-2 and Li-S Batteries. *Electrochim. Acta* **2016**, *192*, 529–550. [CrossRef]
20. Villarreal, J.; Zuniga, L.; Valdez, A.; Alcoutlabi, M. The Use of Mixed Organic/Ionic Liquid Electrolytes with Forcespun Metal Oxides/Carbon Microfiber Electrodes in Lithium Ion Batteries. *ECS Trans.* **2018**, *85*, 387–394. [CrossRef]
21. Garcia, B.; Lavallée, S.; Perron, G.; Michot, C.; Armand, M. Room temperature molten salts as lithium battery electrolyte. *Electrochim. Acta* **2004**, *49*, 4583–4588. [CrossRef]
22. Guerfi, A.; Dontigny, M.; Charest, P.; Petitclerc, M.; Lagace, M.; Vijh, A.; Zaghib, K. Improved electrolytes for Li-ion batteries: Mixtures of ionic liquid and organic electrolyte with enhanced safety and electrochemical performance. *J. Power Sources* **2010**, *195*, 845–852. [CrossRef]
23. Chopade, S.A.; Au, J.G.; Li, Z.; Schmidt, P.W.; Hillmyer, M.A.; Lodge, T.P. Robust Polymer Electrolyte Membranes with High Ambient-Temperature Lithium-Ion Conductivity via Polymerization-Induced Microphase Separation. *ACS Appl. Mater. Interfaces* **2017**, *9*, 14561–14565. [CrossRef]
24. Alarco, P.J.; Abu-Lebdeh, Y.; Abouimrane, A.; Armand, M. The plastic-crystalline phase of succinonitrile as a universal matrix for solid-state ionic conductors. *Nat. Mater.* **2004**, *3*, 476–481. [CrossRef]
25. Das, S.; Bhattacharyya, A.J. Influence of water and thermal history on ion transport in lithium salt-succinonitrile plastic crystalline electrolytes. *Solid State Ionics* **2010**, *181*, 1732–1739. [CrossRef]
26. Wu, X.L.; Xin, S.; Seo, H.H.; Kim, J.; Guo, Y.G.; Lee, J.S. Enhanced Li+ conductivity in PEO-LiBOB polymer electrolytes by using succinonitrile as a plasticizer. *Solid State Ionics* **2011**, *186*, 1–6. [CrossRef]
27. Fan, L.Z.; Hu, Y.S.; Bhattacharyya, A.J.; Maier, J. Succinonitrile as a versatile additive for polymer electrolytes. *Adv. Funct. Mater.* **2007**, *17*, 2800–2807. [CrossRef]
28. Zha, W.P.; Chen, F.; Yang, D.J.; Shen, Q.; Zhang, L.M. High-performance Li6.4La3Zr1.4Ta0.6O12/Poly(ethylene oxide)/Succinonitrile composite electrolyte for solid-state lithium batteries. *J. Power Sources* **2018**, *397*, 87–94. [CrossRef]
29. Chen, R.J.; Liu, F.; Chen, Y.; Ye, Y.S.; Huang, Y.X.; Wu, F.; Li, L. An investigation of functionalized electrolyte using succinonitrile additive for high voltage lithium-ion batteries. *J. Power Sources* **2016**, *306*, 70–77. [CrossRef]
30. Kim, Y.S.; Kim, T.H.; Lee, H.; Song, H.K. Electronegativity-induced enhancement of thermal stability by succinonitrile as an additive for Li ion batteries. *Energy Environ. Sci.* **2011**, *4*, 4038–4045. [CrossRef]
31. Liu, C.F.; Neale, Z.G.; Cao, G.Z. Understanding electrochemical potentials of cathode materials in rechargeable batteries. *Mater. Today* **2016**, *19*, 109–123. [CrossRef]
32. Agubra, V.A.; De la Garza, D.; Gallegos, L.; Alcoutlabi, M. ForceSpinning of polyacrylonitrile for mass production of lithium-ion battery separators. *J. Appl. Polym. Sci.* **2016**, *133*. [CrossRef]
33. Agubra, V.A.; Zuniga, L.; De la Garza, D.; Gallegos, L.; Pokhrel, M.; Alcoutlabi, M. Forcespinning: A new method for the mass production of Sn/C composite nanofiber anodes for lithium ion batteries. *Solid State Ionics* **2016**, *286*, 72–82. [CrossRef]

34. Kim, C.; Park, S.H.; Cho, J.K.; Lee, D.Y.; Park, T.J.; Lee, W.J.; Yang, K.S. Raman spectroscopic evaluation of polyacrylonitrile-based carbon nanofibers prepared by electrospinning. *J. Raman Spectrosc.* **2004**, *35*, 928–933. [CrossRef]
35. Babu, V.S.; Seehra, M.S. Modeling of disorder and X-ray diffraction in coal-based graphitic carbons. *Carbon* **1996**, *34*, 1259–1265. [CrossRef]
36. Flores, D.; Villarreal, J.; Lopez, J.; Alcoutlabi, M. Production of carbon fibers through Forcespinning (R) for use as anode materials in sodium ion batteries. *Mater. Sci. Eng. B* **2018**, *236*, 70–75. [CrossRef]
37. Dahbi, M.; Ghamouss, F.; Tran-Van, F.; Lemordant, D.; Anouti, M. Comparative study of EC/DMC LiTFSI and LiPF6 electrolytes for electrochemical storage. *J. Power Sources* **2011**, *196*, 9743–9750. [CrossRef]
38. Jiang, Y.Y.; Qin, C.D.; Yan, P.F.; Sui, M.L. Origins of capacity and voltage fading of LiCoO2 upon high voltage cycling. *J. Mater. Chem. A* **2019**, *7*, 20824–20831. [CrossRef]
39. Logan, E.R.; Tonita, E.M.; Gering, K.L.; Li, J.; Ma, X.W.; Beaulieu, L.Y.; Dahn, J.R. A Study of the Physical Properties of Li-Ion Battery Electrolytes Containing Esters. *J. Electrochem. Soc.* **2018**, *165*, A21–A30. [CrossRef]
40. Markovsky, B.; Talyossef, Y.; Salitra, G.; Aurbach, D.; Kim, H.J.; Choi, S. Cycling and storage performance at elevated temperatures of LiNi0.5Mn1.5O4 positive electrodes for advanced 5 VLi-ion batteries. *Electrochem. Commun.* **2004**, *6*, 821–826. [CrossRef]
41. Kuhnel, R.S.; Lubke, M.; Winter, M.; Passerini, S.; Balducci, A. Suppression of aluminum current collector corrosion in ionic liquid containing electrolytes. *J. Power Sources* **2012**, *214*, 178–184. [CrossRef]

© 2020 by the authors. Licensee MDPI, Basel, Switzerland. This article is an open access article distributed under the terms and conditions of the Creative Commons Attribution (CC BY) license (http://creativecommons.org/licenses/by/4.0/).

Review

Review of Recent Nuclear Magnetic Resonance Studies of Ion Transport in Polymer Electrolytes

Stephen Munoz [1,2] and Steven Greenbaum [1,2,*]

1 Department of Physics & Astronomy, Hunter College of the City University of New York, New York, NY 10065, USA; mehkie@gmail.com
2 Doctoral Program in Physics, CUNY Graduate Center, New York, NY 10016, USA
* Correspondence: sgreenba@hunter.cuny.edu; Tel.: +01-212-772-4973

Received: 18 September 2018; Accepted: 20 November 2018; Published: 30 November 2018

Abstract: Current and future demands for increasing the energy density of batteries without sacrificing safety has led to intensive worldwide research on all solid state Li-based batteries. Given the physical limitations on inorganic ceramic or glassy solid electrolytes, development of polymer electrolytes continues to be a high priority. This brief review covers several recent alternative approaches to polymer electrolytes based solely on poly(ethylene oxide) (PEO) and the use of nuclear magnetic resonance (NMR) to elucidate structure and ion transport properties in these materials.

Keywords: polymer electrolytes; ion transport; nuclear magnetic resonance (NMR)

1. Introduction

There is an ongoing quest to exploit the full potential energy embodied in the metallic Li^+/Li electrochemical couple in practical and safe battery systems. Using a pure lithium anode material will increase volumetric and mass specific energy density by up to a factor of two while reducing battery cell manufacturing complexity—both key next steps for electrified transportation and consumer electronics [1]. To this end, solid state electrolyte materials have been under investigation for many decades, and the history of polymer-based systems has been with us since the 1970's when polyethylene oxide (PEO) containing alkali metal salts was discovered to be an ionic conductor [2]. Though used for niche applications in thin-film all-solid-state configurations, ceramic or glassy solid electrolytes have recently experienced a strong resurgence in activity, due in part to the discovery of LGPS ($Li_{10}GeP_2S_{12}$) [3] and its Si analogue [4]. Recently, a novel class of glassy electrolytes and electrode reactions has been proposed that work with both lithium and sodium ions [5], though these have been demonstrated in cells with operational potentials under 3 V. The reactive RF (Radio Frequency) sputter deposited LiPON (Lithium Phosphorous OxyNitride) system was demonstrated 20 years ago [6], and has proven to be difficult and costly to scale commercially, even when the useable cell area is on the order of square centimeters or smaller; large area cells are extremely problematic due to the formation pinholes and defects during deposition. Other variations of inorganic electrolyte systems have either been unable to suppress Li dendrites due to Li growth around ceramic grain boundaries have had very low room temperature conductivities, or have been unstable/not demonstrated with high potential cathode systems [7]. For large format applications where uniform thickness and composition over a wide geometric area are of paramount importance, ceramics and glasses electrolyte layers will pose substantial challenge as they suffer from structural rigidity resulting in loss of contact upon repeated cycling.

Although polymers can circumvent some of these issues, there are multiple critical performance parameters that dictate how a solid polymer will function in a battery environment, including ionic transport, mechanical stability, electrochemical stability (at high and low voltage) interfacial integrity, and ability to function at high rate and aerial capacity. Variants of the polyethylene oxide (PEO)-based polymer electrolytes have dominated the academic literature in this field. Approximately 40 years of research on PEO-based polymer electrolytes [8–18] has shown that achieving sufficiently high cationic conductivity (~10^{-4} S·cm^{-2} or better) at room temperature with high voltage cathode materials remains elusive. In the PEO system, the primary conduction mechanism involves the cooperative motion of cations and their coordinating polyether segments, which occurs in the amorphous phase of these often heterogeneous polymer-salt complexes above their glass transition temperature [10]. This has led to decades of effort on suppressing the crystalline phase and lowering the amorphous phase T_g for ambient temperature operation, with only incremental improvement in performance. Angell [19] defined a useful concept, the so-called decoupling index, which parameterizes the degree to which ionic and host structural relaxations are decoupled, and more recently Sokolov [20] recognized that solvent-free polymer electrolytes will probably never achieve a high enough level of conductivity unless the need for this coupling mechanism is eliminated or severely limited. Another consequence of the reliance on polymer segmental motion for ionic conductivity is that the Li$^+$ transference numbers in PEO-based materials tend to be rather low, typically 0.25 or less [13].

Other necessary properties include: stability against lithium metal, ability to fill the material with a high volume percentage of active inorganic materials, swelling and solvent resistance, low electronic conductivity, and ease of processing.

To date, there have been few demonstrated practical device-level results showing the performance and stability of dry polymer electrolytes in functional energy storage devices using a lithium metal anode layer. Many published results elucidate conductivity as a function of temperature, and some assess the chemical stability of the material under anodic and cathodic potentials, though do not include data from full electrochemical cells including rate capability and cycle life studies. There are examples that show the cycling behavior of full cells, though these all have similar characteristics, including at least several of the following: cathode active material areal loading values are significantly lower than those used in practical lithium based batteries, cathode materials having a redox potential below 3.5 V, elevated temperature testing, and very small coin cell or swagelock test format, and all have current densities at 125 mA/cm^2 and below). Table 1 is an accounting some of these results, many of which were recently published in premiere journals.

Table 1. Recent cell-level results reported from cells with dry solid polymer electrolytes. (Courtesy of Prof. Jay Whitacre, Carnegie Mellon University).

Dry Polymer Electrolyte Type	Room Temperature Conductivity Scm^{-1}	Cathode Type Used	Cathode Loading Used (wt. % Active)	Areal Cathode Capacity	Test Fixture Format	Testing Temperature Used	# of Full/Deep Cycles Demonstrated
PEO/nanocomposite [21]	~10^{-5} or lower	LiFePO$_4$ (<3.5 V)	60%	~1 mAh/cm^2	Coin cell	100 °C	100
Polyether/LiFTSI [22]	~8 × 10^{-5}	LiFePO$_4$ (<3.5 V)	54%	Undisclosed	Coin cell	80 °C	1300
PEO/nano particle composite [23]	~5 × 10^{-5}	LiFePO$_4$ (<3.5 V)	63%	Undisclosed	Coin cell	70 °C	130
Single-ion BAB triblock copolymer [24]	Lower than 10^{-6}	LiFePO$_4$ (<3.5 V)	60%	8 mAh/cm^2	Coin cell	80 °C	~100
Block Co-polymer (P3HT-PEO) [25]	~10^{-5} or lower	LiFePO$_4$ (<3.5 V)	50%	Undisclosed	Coin cell	90 °C	10's
Ordered Liquid Crystalline (meogen/Li salt) [26]	~10^{-6} Scm^{-1}	LiFePO$_4$ (<3.5 V)	65%	Undisclosed	Coin cell	60 °C	30
PEO/MEEGE [27]	~ lower than 10^{-5}	LiFePO$_4$ (<3.5 V)	83%	Undisclosed	Pouch Cell	60 °C	250
P(EO/MEEGE/AGE) [28]	lower than 10^{-5}	Nano-coated LiCoO$_2$	82%	~1 mAh/cm^2	Coin cell	60 °C	25 (not fully stable at cathode potentials)
PEM [29]	<10^{-3}	LiFePO$_4$	80%	0.8–1.5 mg/cm^2	Coin cell	ambient	50 cycles (80% capacity after)
Interlinked solid polymer electrolyte [30]	~10^{-4}	LiFePO$_4$ (2.5–4 V)	Undisclosed	~0.1 mAh/cm^2	Coin cell	20 °C	50
Single ion triblock copolymer [31]	<10^{-7}	LiFePO$_4$	60%	Undisclosed	Undisclosed	70 °C	300 (77% capacity retention)
Carbonate-linked PEO electrolyte [32]	<10^{-5}	LiFePO$_4$	80%	1.3–1.8 mAh/cm^2	Coin	25 °C	20

In reviewing the table, several things become apparent: none of these show performance at current densities (>0.5 mA/cm^2) which are needed for practical devices, most are at high temperatures, and none stably incorporate cathode materials of practical importance with technologically relevant loading levels. Nonetheless, the current status of the inorganic sulfide solid electrolytes and the daunting scale-up problems they face, continues to motivate worldwide research into polymer electrolytes.

Ion transport characteristics remain a limiting factor on the practical applicability of many next-generation candidate battery electrolytes. NMR is especially well-suited to studying these properties, as it can easily probe much of the relevant time and length scales, while individually measuring the movement of the various constituents.

The net magnetization of the particles excited in the course of NMR experiments returns to equilibrium according to two relaxation profiles (longitudinal, or T_1, and in-plane, or T_2). Both the value of the relaxation rates and their characters (one- or multi-component) can be determined for those electrolyte components which can be tracked with an NMR-active nucleus. These relaxation rates are determined by intra- and inter-molecular spin interactions, and thus provide insight on the short-range dynamics of the system [33]. In fact, relaxation measurements have been used to probe dynamics since nearly the advent of NMR itself [34].

Long-range dynamics are also suitable for study by using pulsed field gradient NMR methods, which have been in use for the past five decades [35]. As the Larmor frequency of precession is determined by the local magnetic field strength, a magnetic field gradient encodes the position of particles in their phase. Pulsed magnetic field gradients make it possible to encode and decode positions while retaining high signal resolution. The resulting signal intensities can be compared to measure the self-diffusion coefficients, which can be used to calculate several properties germane to battery application.

This is all in addition to the structural characterizations which NMR is well-known for. The ability of NMR to investigate the coordination and solvation of particles in not only liquids, but also solids through the use of magic angle spinning, cross-polarization, and decoupling techniques has been leveraged for many years. More detailed explanations of their application can be found in the literature [36].

The purpose of this review is to examine several recent developments in the literature related to NMR-based investigations of ion transport in selected families of polymer electrolytes, most involving some modification of PEO. Though not exhaustive, we believe that the examples we have chosen to highlight are representative of the majority of current approaches to viable polymer electrolytes, with NMR as a primary analytical tool.

2. PEO and Ceramic Composite Electrolytes

Poly(ethylene oxide) was among the first polymers to be discovered to be an ionic conductor, and the decades since this discovery has seen much time and energy put into reaching its potential as a solid polymer electrolyte. It forms the basis of many more complex polymer systems, including many composites and copolymers [12,37], thanks to the wealth of information on its mechanical and electrochemical properties. Although PEO tends to suffer from low room-temperature conductivity values, the significant advantages it brings in terms of promoting Li salt dissolution, as well as its mechanical properties as a solid polymer, justifies the continued interest in its refinement as an electrolyte.

Polymer/ceramic composite electrolytes are an attractive option for customized mechanical and electrochemical properties. For decades, it has been known that incorporating certain ceramic materials into the polymer matrix can improve the ionic conductivity of the material, mitigating one of the key weaknesses of solid polymer electrolytes [38,39]. This effect is achieved via surface groups of the ceramic particles modifying local structure, as suggested by studies investigating particles of reduced size [40,41]. The particles are believed to affect the recrystallization of the polymer chains, resulting in amorphous regions conducive to fast Li⁺ ion transport [42]. Inclusion of nanowires in lieu of particles can provide a long-range network for improved lithium mobility [43,44]. Ceramic additives enhance the ability of the electrolyte to form a stable interface with electrodes [39,45]. Certain combinations of polymer and ceramic can even result in a greatly increased Li⁺ ion transference number due to cross-linking of the polymer chain promoted by the presence of the ceramic filler, resulting in Li⁺-preferred transport channels near the particles [21].

More recently, "polymer-in-ceramic" electrolytes composites have demonstrated good mechanical properties, high discharge capacities, and good capacity retention in solid state lithium-metal batteries [46]. Incorporation of Li-ion conductive ceramics with a high shear modulus can have the effect of increasing the mechanical resistance to lithium dendrite formation. This, paired with the somewhat Li-insulating nature of the polymer matrix, results in the suppression of dendritic growth while still allowing proper conduction of lithium ions [47]. Thanks to this improved interfacial stability, lithium-metal compatible ceramic composite electrolytes have shown promising behavior [48,49].

A recent study by Zheng et al. [50] focused on elucidating the somewhat complicated nature of Li-ion transport through composite materials, where the ions might transport through the polymer matrix, through the ceramic fillers, and/or through their interfaces. Cubic-$Li_7La_3Zr_2O_{12}$ (LLZO) dry powder was added to a polymer matrix consisting of poly(ethylene oxide) and lithium bis(trifluoromethanesulfonyl)imide (LiTFSI), then ball milled. The resulting slurry was solution cast and dried into a composite film. Several films were cast with different wt. % fractions of LLZO, from 5 wt. % to 50 wt. %. Finally, a separate sample was cast with tetraethylene glycol dimethyl ether (TEGDME) included, at 20 wt. % TEGDME and 50 wt. % LLZO.

^{6}Li solid-state magic-angle spinning NMR was performed to characterize the local structure and dynamics of the lithium ions. The results contain a peak representative of LLZO decomposed through the ball-milling process at 1.3 ppm (relative to LiCl). The results also show a new peak at 1.8 ppm relative to LiCl, indicative of the LLZO–PEO interface [51]. In the sample containing TEGDME, this peak was observed to slightly shift and its area integral to increase. This, along with the increased intensity of the decomposed LLZO peak, confirmed that the TEGDME assists in the breakdown of LLZO, and may play a role in converting more of it to an interfacial complex.

Broadening suggestive of disorder of the local environments for lithium ions is observed in the LiTFSI peak at higher concentrations of LLZO, characteristic of reduced polymer crystallization. A slight reduction in the FWHM in the sample containing TEGDME can be attributed to a partial averaging of the anisotropic interactions due to the increased mobility of the lithium ions. Evidence of this increased mobility was also present in a reduced T_1 of the decomposed LLZO signal for the sample containing TEGDME.

Li-ion transport was further investigated by using ^{6}Li metal electrodes in symmetric cells, and then cycling them to enrich the ^{6}Li in the polymer electrolyte via isotopic exchange. The low natural abundance of the ^{6}Li isotope (7.6%) means that the pathways preferred by Li-ion transport should experience a noticeable enrichment of ^{6}Li (Figure 1).

The results reveal that for the 5 wt. % LLZO sample, an enrichment in the LiTFSI signal is observed, along with a shift in the peak resonance. This change in the ions' electronic environment is consistent with reduced PEO-Li interaction in amorphous phase PEO, leading to faster Li-ion conduction (Figure 1a).

Combined with the T_1 data and CPMAS (^{1}H–^{6}Li) showing very little interaction between LLZO and the PEO matrix, the authors were abler to conclude that the 20% LLZO composite still mainly conducts lithium via the polymer matrix, with the decomposed LLZO assisting the ionic conduction.

The LLZO 50 wt. % sample produced spectra suggesting that the main conduction pathway had changed, with the bulk of enrichment occurring in the LLZO peak, with some in the LiTFSI and interface peaks (Figure 1c). No enrichment was observed to occur in the decomposed LLZO peak. There was now enough LLZO to form a coherent network for the ions to travel through.

Finally, the LLZO 50 wt. % + TEGDME sample spectra revealed that the Li-ion conduction pathway changed again, back to the decomposed LLZO and LiTFSI. This is consistent with TEGDME's high natural ionic conductivity, as well as its ability to reduce PEO crystallization, resulting in preferred movement for lithium through the polymer/TEGDME matrix.

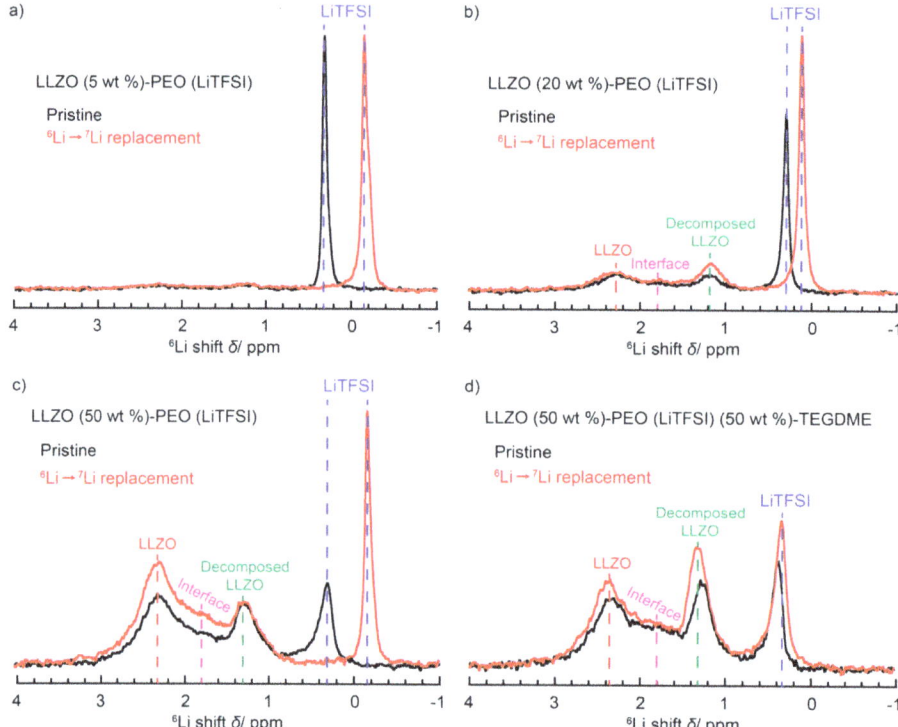

Figure 1. ^{6}Li NMR comparison of pristine and cycled LLZO (5 wt. %)-PEO (LiTFSI), LLZO (20 wt. %)-PEO (LiTFSI), LLZO (50 wt. %)-PEO (LiTFSI), and LLZO (50 wt. %)-PEO (LiTFSI) (50 wt. %)-TEGDME [50]. Reprinted with permission from [50]. Copyright 2018 American Chemical Society.

Further electrochemical measurements via Electrochemical Impedance Spectroscopy (EIS) would reveal that the 50 wt. % LLZO sample demonstrates the lowest conductivity of the samples ($<1 \times 10^{-5}$ S/cm), due to the PEO pathways being blocked and the LLZO network providing poor conductivity on its own. In contrast, the sample containing TEGDME demonstrated a much higher conductivity ($>5 \times 10^{-5}$ S/cm) due to the TEGDME's ability to facilitate ion conduction channels through the PEO. In fact it can be argued that due to these interfacial issues, there is limited advantage to incorporating a highly conducting ceramic over a non-conducting (in the bulk phase) one [7,12,14,21].

Another study by Lago et al. [23] leveraged solid-state NMR to study a plasticized PEO-based Solid Polymer Electrolytes (SPE) containing anions grafted onto ceramic nanoparticles. The idea was to combine the improved conductivity and electrochemical stability of a lithium-only conduction polymer with the increased ionic dissociation, inhibited crystallization, and improved mechanical properties associated with incorporated ceramic nanofillers [45].

Variable-temperature ^{19}F solid-state NMR was performed on two samples: one, the classic PEO(LiTFSI) [EO:Li 20:1], and one composite sample comprised of 5 nm Al_2O_3 ceramic nanoparticles functionalized simultaneously with lithium 4-[2-(trimethoxysilyl)ethyl]benzene-1-sulfonyl [(trifluoromethyl)-sulfonyl]amide and PEG9 trimethoxysilane [EO:Li 50:1] in PEO:PEGDME 1:1 (the choice of nanoparticles was based on a previous work) [52].

Figure 2 shows the comparison of the resultant linewidths of the anion signals in the two samples as a function of inverse temperature. A clear difference in the linewidth response is evident. The linewidth in the LiTFSI–PEO system is heavily dependent on temperature, a consequence of the fact that the mobility of the fluorine in the LiTFSI molecules is coupled to the mobility of the PEO matrix. As the temperature increases, the PEO segments become much less rigid, allowing greatly increased freedom of movement to the LiTFSI molecules, whose molecular tumbling averages out the local anisotropic interactions and results in a much-narrowed NMR peak. To the contrary, the relative temperature-independence of the sample containing functionalized Al_2O_3 nanoparticles indicates that the local mobility of the anions is decoupled from that of the polymer matrix. Furthermore, the larger linewidths in the sample containing nanoparticles at higher temperatures confirms that, although their movement is decoupled from that of the polymer, it does experience restriction due to its association with the Al_2O_3 nanoparticles.

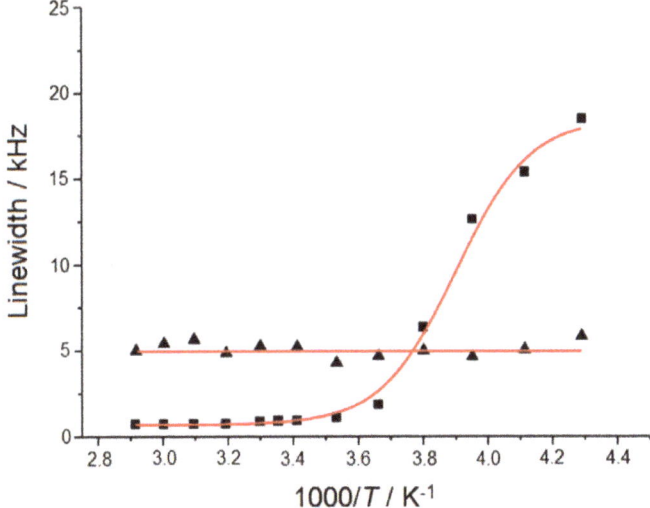

Figure 2. Comparison of ^{19}F NMR linewidths for (triangle) Al_2O_3-PEG9-anion (5 nm). [EO]/[Li]~50 in PEO/PEGDME~1, and (square) LiTFSI in PEO [EO]/[Li]~20 as a function of temperature [23] © 2015 Wiley-VCH Verlag GmbH & Co. KGaA, Weinheim.

EIS measurements would reveal that this composite material shows conductivity approaching 10^{-4} S/cm at 70 °C. This electrolyte was then used to create an Li-metal/LiFePO$_4$ cell which demonstrated better cycling performance than previous Li-metal batteries with composite polymer electrolytes [53]. This, combined with the respectable conductivity and high cation transference resulting from the immobilization of anions, means that this approach could represent a viable path forward on the development of a practical solid-state battery.

These results demonstrate the ability of solid-state NMR to discern the different contributors to ion transport in these complex materials. Polymer/ceramic composites represent a polymer electrolyte family with excellent potential, thanks to its mechanical and electrochemical customizability and compatibility. NMR can be instrumental in developing models to guide future design of these promising electrolyte candidates, or in verifying critical aspects of their performance.

3. Copolymers, Block Copolymers, and Polymer Blends

There is significant interest in the use of copolymers as battery electrolytes, due to the fact that different components can be used to selectively engineer the nanostructure, theoretically leading to advantageous macroscale properties [54,55]. Crystallization of polymer-based electrolytes has been shown to limit the conductivity below practical application levels [56], but incorporating copolymers has been shown to be a viable way to mitigate this crystallinity [57–59]. Phase separation can assist in both improving conductivity and in inhibiting lithium dendrite growth through mechanical rigidity.

A recent study by Daigle et al. investigated the Li$^+$ ion mobility in comb-like copolymers via solid-state NMR [60]. These comb-like polymers were based on poly(styrene) (PS) backbone fashioned through anionic polymerization. The purpose of this backbone was to provide mechanical reinforcement to inhibit lithium dendrite growth via the phenyl groups. Poly(ethylene glycol) methyl ether methacrylate (PEGMA, shown in Figure 3) was grafted to assist in Li$^+$ conductivity by suppressing crystallinity. LiTFSI salt was incorporated to provide charge carriers. Several samples were created with differing ratios of PEGMA to PS (2.6:1, 3.9:1, and 30:1). Solid-state cross-polarization (CP) and direct acquisition ^{13}C NMR measurements were performed to characterize the structure of the polymers, while ^{7}Li NMR measurements were performed to track the li-ion transport mechanics.

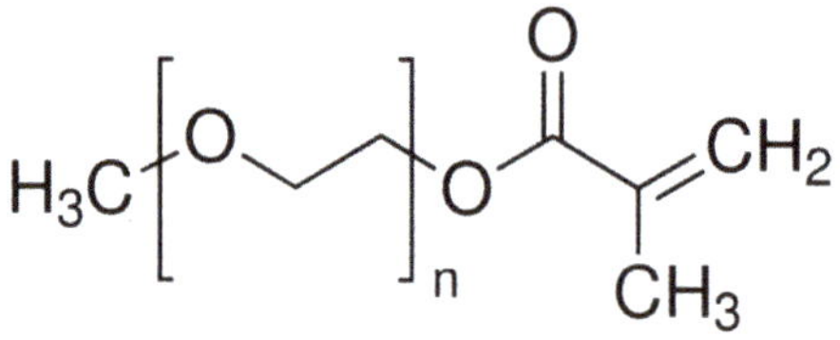

Figure 3. Poly(ethylene glycol) methyl ether methacrylate (PEGMA) chemical structure.

The structural analysis revealed that, as expected, the PEGMA backbones were more rigid than the pendant groups. However, a signal attributed to the pendant groups was acquired in the CP measurements; due to the fact that some rigidity is necessary to facilitate the magnetization transfer necessary for a CP measurement, the authors concluded that the coordination between the pendant groups and lithium salts resulted in this rigidity.

^{7}Li NMR was then used to elucidate the Li-ion transport mechanisms. The samples with the lower ratios of PEGMA to PS displayed conductivities above 10^{-4} S/cm at 60 °C. This approaches what could be considered high enough conductivity for practical application. This suggests the potential for these materials for use in energy storage, and the importance of understanding the mechanisms underlying their operation.

Lithium ion diffusion was deduced by examining the linewidths of the peaks produced in the spectra (it is possible to relate these linewidths to transverse relaxation of the signal, mediated by short-range interactions). Similarly, short-range motion can be correlated with longitudinal relaxation times, also measurable via NMR.

The linewidths, measured across the three samples as a function of temperature, reveal that the lithium signals produce very sharp peaks when compared to copolymers based on polyurethane-poly(dimethylsiloxane) [61]. This is correlated with more mobility of the ions, which is corroborated by the fact that conductivity (measured here by AC impedance spectroscopy) is several times higher than in that previous material, in the case of the samples with lower PEGMA/PS ratio. In addition, when ^{1}H decoupling was applied, no change was observed in the signal, leading the authors to conclude that the lithium-ion mobility was high enough to motionally average out ^{1}H–^{7}Li dipolar interactions.

T_1 as a function of temperature, shown in Figure 4, would reveal that sample 2 (PEGMA:PS 3.9:1) demonstrates the highest lithium mobility, while at the same time showing a weaker temperature dependence than the other two samples. However, all three samples demonstrate a sharp drop in mobility around 263 K.

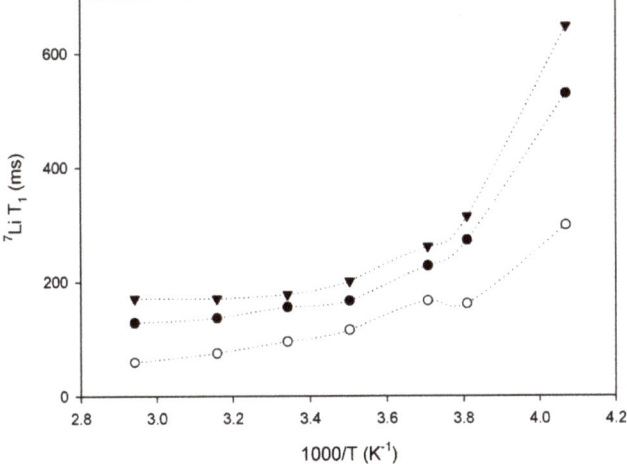

Figure 4. Evolution of ^{7}Li linewidth as a function of inverse temperature for sample 1 (**black triangles**), sample 2 (**open circles**) and sample 3 (**black circles**) [60] (OPEN ACCESS).

The similarity between the lithium ion mobility and PEGMA chain mobility allowed the authors to conclude that their movement is correlated.

Another recent study reported on eight PEO-polycarbonates [62]. This study was motivated by research showing that aliphatic polycarbonates could enable room-temperature cycling [63,64]. The authors prepared several different samples of PEO-PC polymer, varying both the ratio of PEO to PC and the LiTFSI salt concentration. ^{1}H, ^{13}C, ^{19}F, and ^{7}Li NMR experiments were performed to characterize the structure and local dynamics of the system.

The authors elected to investigate the effect of varying salt concentration on the PEO-PC (34:1) sample, owing to it displaying the highest room temperature conductivity as measured by AC impedance spectroscopy across the entire temperature range measured. ^{7}Li relaxation experiments would reveal that a new signal appears for both the ^{7}Li and ^{19}F spectra at the highest concentration of salt, 80 wt. % LiTFSI (despite significant shimming issues affecting the lineshapes, the authors note a discernable difference between the attenuation of the two peaks, concluding that the secondary peak is not an artifact of shimming. However, it should be noted that the very broad lineshapes of the secondary peaks can affect the accuracy of any relaxation times derived thereof.) The authors ascribe this secondary peak to the formation of LiTFSI aggregates. The spectra are displayed in Figure 5.

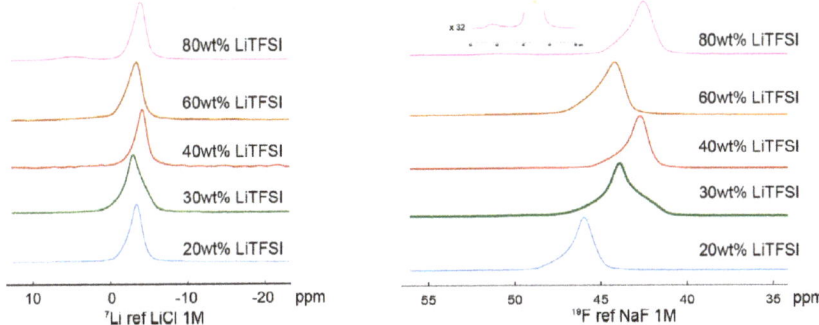

Figure 5. (**left**) ^{7}Li NMR spectrum of PEO34-PC at 343.15 K of different wt. % LiTFSI samples. (**right**) ^{19}F NMR spectrum of PEO34-PC at 343.15 K of different wt. % LiTFSI samples [62] Reprinted from [62]. Copyright 2018, with permission from Elsevier.

The BPP (Bloembergen, Purcell, and Pound) model [34] was fitted with the resulting T_1 measurements to calculate the correlation time, activation energy, and quadrupolar coupling constant for the samples of varying salt concentration. The activation energy was found to drop dramatically with higher salt concentrations, consistent with faster reorientational dynamics. Higher correlation times were calculated for the ion aggregates, consistent with slower dynamics and less mobility.

^{7}Li and ^{19}F pulsed field gradient NMR was also performed on the PEO-PC (34:1) samples with differing salt concentration. Of note is a sharp uptick in the diffusion coefficients of both ^{7}Li and ^{19}F at 80% wt. LiTFSI, which the authors note contradicts the AC impedance spectroscopy-measured conductivity values which follow a consistent downward trend with higher salt concentration. This is explained through the fact that in many systems, the diffusion can be so slow, or the relaxation so fast, that certain species may go undetected in the course of the experiment. It is very likely that the Li ion's share of the lithium signal is dying out before ever being acquired by the spectrometer, resulting in diffusion coefficients being calculated from the attenuation of just a tiny fraction of the "true" signal. This is an illustration of the importance of tempering conclusions made from solely NMR by comparing against other methods of measurement.

Another recent study, carried out by Timachova et al. [65], focused on a nanostructured block copolymer electrolyte. These electrolytes are of interest because they can form nanoscaled ordered regions of alternating phase, which enables the kind of combinations of rigidity and conductivity necessary in a practical battery electrolyte. Much work has focused on the characterization of these materials due to their attractiveness as electrolytes [66,67]. In 2016, Chintapalli et al. [68] reported a study of polystyrene block poly(ethylene oxide) (SEO) mixed with LiTFSI. Through a combination of differential scanning calorimetry (DSC) and AC impedance spectroscopy, they determined that the maximum conductivity of the SEO occurred at very different salt concentrations than in the PEO ($r = 0.21$ as opposed to $r = 0.11$). This is due to inhibited grain growth, which increases the ionic conductivity of the block copolymer.

Timachova et al. applied PFG-NMR to study a polystyrene-b-poly (ethylene oxide) copolymer/ lithium bis(trifluoromethanesulfonyl)imide solid electrolyte as a function of salt concentration. Their goal was to characterize the local anisotropic nature of Li-ion diffusion due to the lamellar layers, and to obtain the isotropic continuum transport properties (a first for block copolymer electrolytes).

The SEO in the samples was synthesized via sequential anionic polymerization of styrene followed by ethylene oxide [69]. Electrolytes of several different LiTFSI concentrations were then prepared ($r = $ [Li]/[EO] = 0.03, 0.06, 0.12, 0.18, 0.24, and 0.3).

Pulsed-field gradient NMR was performed on the electrolyte samples, targeting both ^{7}Li and ^{19}F nuclei to track the movement of cations and anions. Initial comparison of the attenuation curves of a traditional PEO electrolyte with that of the SEO block copolymer reveals a clear difference, as seen in Figure 6.

A linear relationship between the normalized signal intensity and the square of the gradient strength is indicative of isotropic diffusion in this case, as illustrated by the PEO result. In contrast, the SEO electrolyte produced a nonlinear relationship, indicating anisotropic diffusion. The curve through Figure 6b represents the best fit of the anisotropic diffusion coefficient with D∥ (diffusion along the lamellae) and D⊥ (diffusion perpendicular to the lamellae).

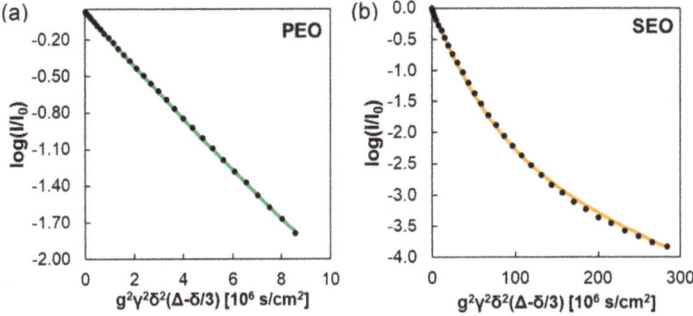

Figure 6. PFG-NMR signal attenuation of ^{19}F seen in (**a**) PEO(5)/LiTFSI at $r = 0.06$ and (**b**) SEO(16−16)/LiTFSI at $r = 0.18$ [65]. Reprinted with permission from [65]. Copyright 2018 American Chemical Society.

Figure 7 shows the resulting values through the PEO-rich lamellae (associated with D∥) and across the PEO/poly(styrene) boundaries (associated with D⊥). AC impedance spectroscopy was performed to measure the conductivity, and steady-state current and restricted diffusion measurements were performed to help calculate the steady-state transference number. These values were then used to calculate the Stefan-Maxwell diffusion coefficients, represented by the light squares in the plots. The calculated diffusion coefficients representing the Li-PEO and TFSI−PEO interactions are consistent with the D⊥ values measured by PFG-NMR. This strong agreement from two different measurements allowed the authors to conclude that the electrochemical performance is strongly coupled with ion transport through defects in this system.

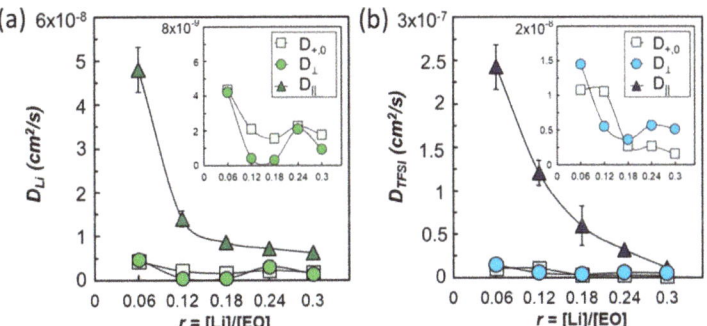

Figure 7. Parallel, D∥, and perpendicular, D⊥, diffusion coefficients and the Stefan−Maxwell diffusivities of (**a**) Li and (**b**) TFSI in SEO(16−16) as a function of salt concentration, r, at 90 °C [65]. Reprinted with permission from [65]. Copyright 2018 American Chemical Society.

This result is significant because it represents the first time that the local anisotropic nature of diffusion has been characterized in a block copolymer electrolyte. The authors would leverage this new information to establish an NMR-based "morphology factor" representing the degree of isotropy of diffusion. They would find that this factor indicates low isotropy at low concentrations of salt (close to that expected for an ideal lamellar system with D⊥ = 0), and high isotropy with higher concentrations of salt. Transmission electron microscopy (Figure 8) would confirm that the lamellar structure contains significantly more defects with higher salt concentration. This provides further evidence that the NMR measurements were able to accurately decouple the in-plane and through-plane diffusion values. The results indicate that transport in the bulk is strongly influenced by defects in the structure, providing guidance for further optimization of these materials.

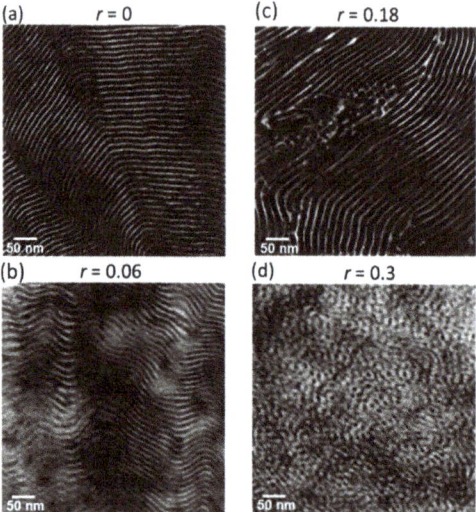

Figure 8. Dark-field transmission electron microscopy images of SEO(16−16) at (**a**) $r = 0$ and (**c**) $r = 0.18$, reproduced from [68] and at (**b**) $r = 0.06$ and (**d**) $r = 0.3$ measured in this work. The bright phase is poly(ethylene oxide) [65]. Reprinted with permission from [65]. Copyright 2018 American Chemical Society.

Another study by Liu et al., in 2017 [70], would focus on a blended hybrid solid polymer electrolyte. Blended polymers are attractive due to the ease of synthesis while still providing a high degree of control over the mechanical properties of the end product [71–73]. This particular material studied was created by blending two organic-inorganic hybrids. One hybrid consisted of (3-glycidyloxypropyl)trimethoxysilane (GLYMO), an organosilane, cross-linked with a monoamine-based polyether (Jeffamine M-2070); the second consisted of GLYMO and poly(ethylene glycol) diglycidyl ether (PEGDGE) reacted with a diamine-based polyether (Jeffamine ED2003). These structures are illustrated in Figure 9. These hybrids would be combined in different ratios and LiClO$_4$ salt added to create several electrolyte samples for examination. Solid-state NMR was performed on ^{13}C, ^{29}Si, and ^{7}Li nuclei to characterize the lithium-ion mobility and confirm the structure of the material.

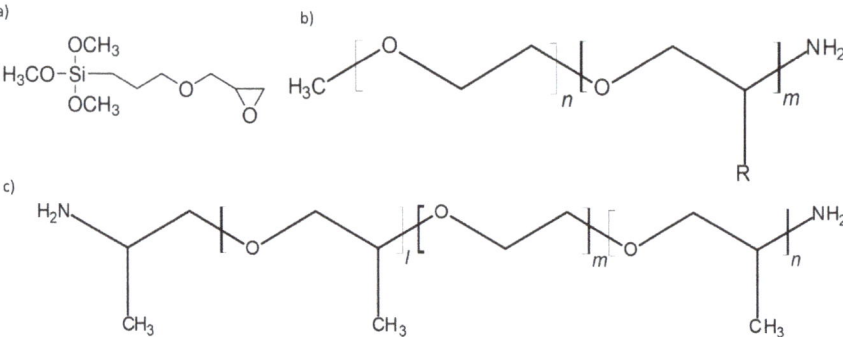

Figure 9. Structures of (**a**) (3-glycidyloxypropyl)trimethoxysilane (GLYMO), (**b**) Jeffamine M-2070, and (**c**) Jeffamine ED2003.

AC impedance measurements would suggest that the ion transport was linked to the segmental motion of the polymer matrix, as is common in many solid polymer electrolytes. The authors determined that a peak ionic conductivity should occur at a salt concentration of [O]/[Li] = 16. This was based on the fact that increasing the salt concentration can have competing effects on conductivity—higher concentration increases the amount of charge carriers, but too high of a concentration can lead to ion aggregation and impede mobility. It should be noted here that some recent studies have shown that certain salt concentrations outside the assumed window of interest can yield competitive performance via interionic interactions [74].

^{7}Li static linewidths were measured across a range of temperatures from −90 °C to 90 °C. Measurements carried out without proton decoupling would be characterized by a plateau of broad linewidths ~5 kHz at temperatures below −60 °C, and another flat region of narrow linewidths above 60 °C. Activation energies were calculated from these results to be 0.15 eV for the sample synthesized from hybrid 1:hybrid 2 (70:30) with [O]/[Li] = 32 (denoted in the article as MP(70:30)-32 and 0.14 eV for MP(70:30)-16, which is comparable to that of similar systems in the literature [75]. With proton decoupling applied, a sharp decrease in the linewidth at lower temperatures was observed, leading the authors to conclude that about 80% of the interaction causing broadening was due to ^{7}Li–^{1}H dipolar interactions. VTF fitting of conductivity measured by AC impedance spectroscopy would confirm that the ionic conductivity is strongly coupled to the polymer segmental chain motion.

Following up on this, ^{1}H-decoupled ^{7}Li MAS measurements were performed in the same temperature range. These measurements revealed that the coordination between Li-ions and the ether oxygens in the polyethers in M-2070, ED2003, and PEGDGE produced the strongest signal, while that of the ^{7}Li coordinated with GLYMO oxygen was extremely difficult to detect. The authors ascribe this lack of signal to low GLYMO concentration in the sample.

Although the MP(70:30)-16 sample showed the highest room temperature conductivity value at over 1x10^{-4} S/cm and an electrochemical stability window approaching 5 V, test cells incorporating it showed significant irreversible capacity loss upon cycling due to the suspected formation of a passivation layer. However, examination after cycling would show that the membrane showed no signs of mechanical decomposition or particle aggregation.

4. Crystalline Polymer Electrolytes

Acceptable conductivity cannot occur in crystalline polymer electrolytes unless the ionic motion can be effectively decoupled from the polymer matrix [76]. In fact, this decoupling is believed to be an important step toward developing polymer electrolytes of any type that can provide the ionic conductivities necessary for practical application [20]. In 2001, Gadjournova et al. showed how crystallinity can be a boon for cation transport, due to the regularity of ordered diffusion pathways [76]. A number of approaches have been suggested in recent years for enhancing the conductivity, including anionic doping, replacement of the ends of the polymer chains with glymes to enhance disorder [77–79], and stretching of the polymer to align the chains and enhance transport along the longitudinal axis [80,81].

Recently, Yan et al. reported a new crystalline solid polymer electrolyte consisting of a PEO-urea-LiTFSI complex [82]. Their investigation was motivated by a previous study on a similar ternary structure involving α-cyclodextrin, which demonstrated fast Li-ion movement through the resultant structure. Urea was chosen as the next candidate, thanks to the formation of a crystalline inclusion compound in PEO-urea binaries [83–85].

The ternary complex was prepared by dissolution of PEO, urea, and LiTFSI in acetonitrile (with varying concentrations of the LiTFSI salt), followed by stirring, casting at 40 °C, and drying. Wide angle x-ray spectroscopy was used to verify the crystalline nature of the resulting compounds. The α-PEO-urea-LiTFSI complex forms the same crystalline structure as α-PEO-urea, only slightly more compact. A high level of crystallinity is maintained. The highest-conductivity samples were investigated via solid-state ^{1}H–^{7}Li and ^{1}H–^{19}F cross-polarization MAS NMR to determine Li$^+$ and TFSI$^-$ ion coordination.

These tests would show evidence of correlation between Li$^+$ ions and both NH$_2$ (at ~6 ppm) in urea and CH$_2$ (at ~4 ppm) in PEO. This suggests that Li$^+$ is present in the crystalline inclusion structure, where both the urea and PEO are known to be present. The same trend is noted for the ^{1}H–^{19}F tests; TFSI$^-$ anions are correlated with both the urea and PEO, suggesting their presence in the hexagonal urea channel as well. Of note here is the presence of two peaks in the ^{19}F spectrum, which the authors ascribe to differing environments. From an NMR perspective, relaxometry studies, as well as diffusometry, could be of use here in further elucidating differences between the environments (however, specialized systems would be required given the relatively slow diffusion expected from their reported conductivities).

The authors describe the resultant system as consisting of the aforementioned inclusion structure consisting of urea channels containing PEO chains and Li$^+$ and TFSI$^-$ ions. There is evidence that urea promotes ionic dissociation between Li$^+$ and TFSI$^-$ [86]. In addition, the authors hypothesize that the channels may trap the larger TFSI$^-$ anions, allowing the Li$^+$ ions to travel freely. This would produce conductivity and transference numbers more favorable for battery operation.

Impedance spectroscopy was also used to determine the conductivities of the sample. The highest conductivity material demonstrates a conductivity of ~6×10^{-5} S/cm at 30 °C. This value compares favorably to previous highly crystalline polymer electrolytes, and even to some amorphous species, but it is still orders of magnitude below what would be required for a commercially viable battery. However, this study demonstrates that exploitation of inclusion complexes similar to these could facilitate competitive conductivities and transference numbers.

In another study, Fu et al. [87] reported a crystalline polymer electrolyte based on self-assembled α-cyclodextrin (CD), polyethylene oxide (PEO), and Li$^+$ salts. Through the literature, they identified two key goals to increasing the ionic conductivity: to decouple the Li$^+$ ions from the polymer chain segments [24,88–96] and to generate long-range pathways for bulk ionic transport [97]. Along these lines, they combined their self-assembly approach to creating ordered nano-tunnels with alteration of the conformational sequence of PEO to inhibit interaction between Li$^+$ ions and the polymer segments. Samples of differing α-cyclodextrin to PEO ratios were synthesized following previous works [98,99]. ^{1}H–^{13}C CP MAS NMR and wide angle X-ray diffraction measurements were performed to verify the tunnel structure of the PEO chains. The ^{13}C NMR spectra are displayed in Figure 10.

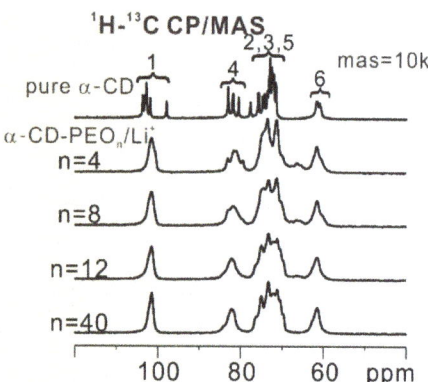

Figure 10. The disappearance of the signal splitting indicates the formation of α-CD-PEO inclusion complex [87] © 2018 WILEY-VCH Verlag GmbH & Co. KGaA, Weinheim.

The NMR results verify that the splitting is no longer resolved in the complex samples compared to the neat α-CD, which has been associated in the literature with the formation of inclusion complexes [100–102].

Static ^{19}F NMR measurements would reveal that the lineshape of the signal changes very little over the temperature range 0–40 °C, with a broadness indicative of low mobility. In contrast, the ^{7}Li static NMR measurement (spectra shown in left of Figure 11) would manifest a narrow signal associated with high mobility. This narrow signal at −1.15 ppm would be designated Li-2 by the authors, associated with Li$^+$ ions between polymer chains and CDs. Broader signals at −1.31 ppm and −0.64 ppm would be assigned to Li$^+$ ions strongly associated with the polymer chains and with the assembled CDs, respectively.

In addition, the narrow Li-2 signal broadened significantly with lower temperature. This indicates that the motion of the Li cations is decoupled from that of the anions, at least in the higher temperature regime. This has important implications for the applicability of this material, as ion pairing can significantly attenuate the practical conductivity of an electrolyte.

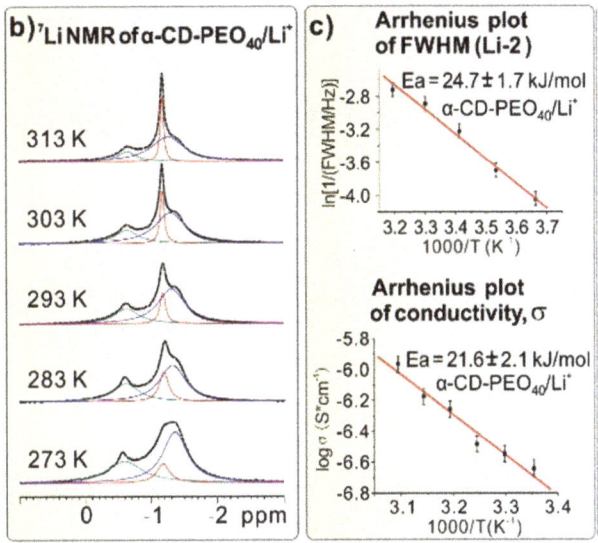

Figure 11. (**Left**) Temperature-dependent ^{7}Li NMR spectra of α-CD-PEO40/Li$^+$ from 273 K to 313 K. (**right**) Top: The Arrhenius plot of the line widths of Li-2 from the temperature-dependent ^{7}Li NMR; Bottom: The Arrhenius plot of the conductivities of α-CD-PEO40/Li$^+$ [87] © 2018 WILEY-VCH Verlag GmbH & Co. KGaA, Weinheim.

Based on the linewidth of Li-2 observed in the NMR measurements, plotted in right of Figure 11, the activation energy was calculated to be about 21.6 kJ/mol, which agrees, within error, with the activation energy calculated from the temperature dependence of the conductivity (also shown in right of Figure 11) measured by EIS (although the authors are careful to mention that this could be coincidental). This low value indicates that the motion of the Li$^+$ ions is coupled much less strongly than in systems such as oxygen-based superionic conductors or EO/Li$^+$ complex crystals [103,104]. This decoupling, in tandem with tunnel structure providing long-range Li$^+$ transport pathways, produces a high EIS-measured conductivity on the order of 1×10^{-3} S/cm at room temperature in the sample with the lowest lithium concentration.

Further ^{2}H solid state NMR measurements would provide evidence that the PEO chains were forming all-trans conformation sequences. Simulations show that no stable structure exists for Li$^+$ ions to coordinate with the PEO chains in such a conformation sequence, further reinforcing the idea that the Li$^+$ ions are very weakly coupled to the PEO chains in this system. In the PEO$_4$ sample, the PEO chains are more likely to conform in the trans-trans-gauche sequence, resulting in stronger coupling between ions and the polymer matrix, with resultant lower ion mobility and conductivity. These results have important implications for the refinement of solid polymer electrolytes which facilitate fast Li$^+$ ion transport, both through the engineering of nanostructure and through attenuation of the interaction between ions and the polymer.

5. Sodium-Conducting Electrolytes

Sodium chemistries represent an attractive alternative to the established lithium-based technology, due to similar properties and much higher abundance. They are limited by a lack of compatible electrodes, which are in turn limited by the need for a suitable electrolyte. A recent study by Pope et al. reports an investigation of a single-ion conducting Na electrolyte [105]. Single-ion conductors promote facile transport of one ion while trapping the counter ion. With proper engineering this can lead to extremely high transference numbers and result in optimized battery performance.

This particular study is motivated by previous works identifying poly(2-acrylamido-2-methyl-1-propane-sulfonate) (PAMPS) as a suitable base for such an electrolyte, thanks to reduced cation tethering to the immobilized anions [106–108].

Further studies have shown that incorporation of a bulky IL quaternary ammonium cations can reduce the T_g by inhibiting crosslinking, leading to increased conductivity [109,110] and have indicated that ether group-containing additives can partially solvate Na$^+$ ions, further assisting in their transport [111]. Putting these ideas together, the authors created an electrolyte by combining a PAMPS homopolymer with an ether group-functionalized quaternary ammonium ion as illustrated in Figure 12. They performed solid-state ^{23}Na NMR experiments to further investigate the Na$^+$ transport mechanism.

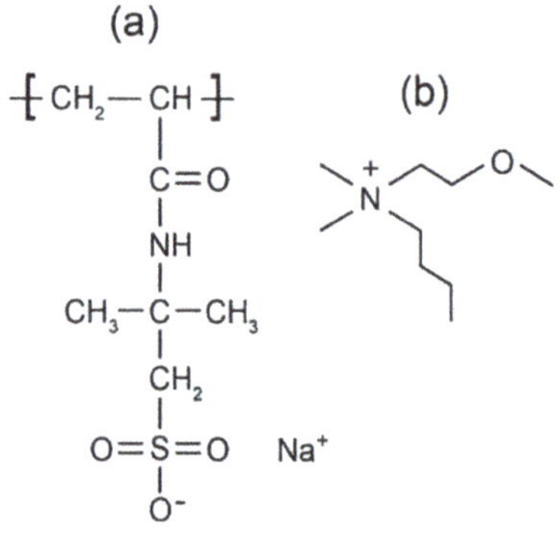

Figure 12. Structure of the ionomer. (**a**) The poly(2-acrylamido-2-methyl-1-propane-sulfonic acid) (PAMPS) monomer and (**b**) the dimethylbutylmethoxyethyl ammonium cation (N114(201)).

Two-component T_1 relaxation results would show strong evidence of at least two different Na populations. The larger-linewidth component was assigned to a less-mobile population, and corresponded to a larger T_1. The difference in linewidths implies an expected difference in T_2 as well, making an exact population split calculation unfeasible due to the differing attenuation effects. Variable-temperature linewidth measurements would reveal that the relaxation times of the different populations become more and more similar above the glass-transition temperature. The authors concluded that the less mobile Na signal can be attributed to ions bound to anionic sulfate groups, with the more mobile signal attributed to unbound Na$^+$ ions.

Despite the T_g-lowering effect of the ammonium cations, the glass-transition temperature remains too high in this system for practical room-temperature application. Regardless, this study demonstrates the selective ability of NMR to decouple the mobilities of different species, even in multi-cationic systems.

Another recent study focuses on the use of electrospinning to fabricate a sodium-ion conducting PEO-based membrane [112]. Sodium is an attractive alternative to lithium-based chemistries, thanks to its abundance and cost advantages. Recent research has shown that electrospinning can produce membranes with enhanced performance when compared to traditional solution-cast membranes [113]. Samples were created by combining PEO with $NaBF_4$ and succinonitrile (SN) in different ratios.

AC impedance spectroscopy would reveal that, near room temperature, the best conductivity was observed in the sample containing ($PEO:SN:NaBF_4$) in the ratio (18:0:1)—indicating that the inclusion of the succinonitrile plasticizer did not provide an improvement in conductivity.

^{19}F and ^{23}Na solid-state NMR was then performed at ~265 K to investigate the local environments. The results would indicate significant mobility differences between the samples of different concentrations. In the case of the ^{19}F spectra, a very broad linewidth is present in the (18:0:1) sample, indicating relative immobility of the BF_4 anions. In contrast, the (18:3:1) sample produces significantly sharper linewidths, consistent with a much higher percentage of mobile anions. ^{23}Na NMR spectra would show the same trends across the different sample concentrations, with the (18:0:1) sample having the lowest cation mobility. Variable temperature investigations of the linewidth of the narrow component of the signals enabled the authors to estimate activation energies of about 42 kJ/mol for Na^+ and BF_4^- ions in the (18:0:1) sample, compared to 39 kJ/mol and 38 kJ/mol for Na^+ and BF_4^-, respectively, in the 18:3:1 membrane. Despite the lack of measurable improvement in conductivity, the authors conclude that the succinonitrile has a significant effect on the ionic mobility in the membranes. In fact, the NMR lineshapes indicated multiple phases in the 18:3:1 sample—one immobile $PEO:NaBF_4$ phase, and one more mobile $PEO:SN:NaBF_4$ phase.

The system was further elucidated by ^{13}C MAS NMR on the 18:3:1 system, which allowed the authors to estimate that about 1/3 of the PEO was in the mobile "SN-activated phase" after deconvolution of the broad and narrow PEO signal components. CPMAS measurements and J-coupling observed with the ^1H decoupling deactivated would provide further evidence of this biphasic behavior. This is also not observed in similar Li-based membranes studied previously [113]. The presence of the immobile phase is thought to be the reason that the observed conductivity in the 18:3:1 membrane is less than that of the 18:0:1 sample, despite the significantly higher local ionic mobility. Pulsed-field gradient NMR may be useful in this case to establish a measurement to combine with the measured conductivity and reveal more information about the long-range dynamics of the system. However, due to the substantial nuclear quadrupole interactions resulting in rapid transverse relaxation associated with ^{23}Na, PFG NMR would be a very challenging undertaking.

6. Conclusions

Nuclear magnetic resonance has proven to be an invaluable tool in the characterization of both structure and dynamics of a wide variety of materials. Its suitability for examining those properties associated with battery performance justifies its continued use in optimizing the electrolytes of the future; advances in both its technology and methodology allow it to remain relevant in the study of the ever-more complex electrolyte systems being developed. The quest for a practical room-temperature solid electrolyte continues, and the polymer electrolyte families described herein are but a small sample of the research towards viability.

Author Contributions: As this is a review article, the primary literature search was conducted by S.M. who also wrote most of the manuscript. S.G. defined the scope of the literature to be reviewed and provided some of the technical content, guidance, and editing.

Funding: The program on Li batteries at Hunter College is supported by the U.S. Office of Naval Research. S.M. acknowledges financial support by the National Institutes of Health RISE program at Hunter College (grant GM060665).

Conflicts of Interest: The authors declare no conflicts of interest.

References

1. Quartarone, E.; Mustarelli, P. Electrolytes for solid-state lithium rechargeable batteries: Recent advances and perspectives. *Chem. Soc. Rev.* **2011**, *40*, 2525–2540. [CrossRef] [PubMed]
2. MacCallum, E.J.R.; Vincent, C.A. *Polymer Electrolyte Reviews*; Elsevier Applied Science: London, UK, 1989.
3. Kamaya, N.; Homma, K.; Yamakawa, Y.; Hirayama, M.; Kanno, R.; Yonemura, M.; Kamiyama, T.; Kato, Y.; Hama, S.; Kawamoto, K.; et al. A lithium superionic conductor. *Nat. Mater.* **2011**, *10*, 682–686. [CrossRef] [PubMed]
4. Bron, P.; Johansson, S.; Zick, K.; Schmedt auf der Günne, J.; Dehnen, S.; Roling, B. $Li_{10}SnP_2S_{12}$: An Affordable Lithium Superionic Conductor. *J. Am. Chem. Soc.* **2013**, *135*, 15694–15697. [CrossRef] [PubMed]
5. Braga, M.; Grundish, N.; Murchison, A.; Goodenough, J. Alternative strategy for a safe rechargeable battery. *Energy Environ. Sci.* **2017**, *10*, 331–336. [CrossRef]
6. Yu, X.; Bates, J.; Jellison, G.; Hart, F. A Stable Thin—Film Lithium Electrolyte: Lithium Phosphorus Oxynitride. *J. Electrochem. Soc.* **1997**, *144*, 524–532. [CrossRef]
7. Fergus, J.W. Ceramic and polymeric solid electrolytes for lithium-ion batteries. *J. Power Sources* **2010**, *195*, 4554–4569. [CrossRef]
8. Fenton, D.E.; Parker, J.M.; Wright, P.V. Complexes of alkali metal ions with poly(ethylene oxide). *Polymer* **1973**, *14*, 589. [CrossRef]
9. Armand, M.B.; Chabagno, J.M.; Duclot, M.J. Fast Ion Transport in Solids. In Proceedings of the International Conference on Fast Ion Transport in Solids, Electrodes and Electrolytes., Lake Geneva, WI, USA, 21–25 May 1979; p. 131.
10. Berthier, C.; Gorecki, W.; Minier, M.; Armand, M.B.; Chabagno, J.M.; Rigaud, P. Microscopic investigation of ionic conductivity in alkali metal salts-poly(ethylene oxide) adducts. *Solid State Ionics* **1983**, *11*, 91–95. [CrossRef]
11. Adamić, K.J.; Greenbaum, S.G.; Wintersgill, M.C.; Fontanella, J.J. Ionic conductivity in solid, crosslinked dimethylsiloxane-ethylene oxide copolymer networks containing sodium. *J. Appl. Phys.* **1986**, *60*, 1342–1345. [CrossRef]
12. Chung, S.H.; Wang, Y.; Persi, L.; Croce, F.; Greenbaum, S.G.; Scrosati, B.; Plichta, E. Enhancement of ion transport in polymer electrolytes by addition of nanoscale inorganic oxides. *J. Power Sources* **2001**, *97–98*, 644–648. [CrossRef]
13. Armand, M.; Tarascon, J.-M. Building better batteries. *Nature* **2008**, *451*, 652–657. [CrossRef] [PubMed]
14. Armand, M.B.; Bruce, P.G.; Forsyth, M.; Scrosati, B.; Wieczorek, W. Polymer Electrolytes. In *Energy Materials*; Bruce, D.W., Dermot, O.H., Wlton, R.I., Eds.; John Wiley & Sons, Ltd.: Hoboken, NJ, USA, 2011.
15. Kovarsky, R.; Golodnitsky, D.; Peled, E.; Khatun, S.; Stallworth, P.E.; Greenbaum, S.; Greenbaum, A. Conductivity enhancement induced by casting of polymer electrolytes under a magnetic field. *Electrochimica Acta* **2011**, *57*, 27–35. [CrossRef]
16. Wetjen, M.; Kim, G.-T.; Joost, M.; Winter, M.; Passerini, S. Temperature dependence of electrochemical properties of cross-linked poly(ethylene oxide)–lithium *bis*(trifluoromethanesulfonyl)imide–*N*-butyl-*N*-methylpyrrolidinium *bis*(trifluoromethanesulfonyl)imide solid polymer electrolytes for lithium batteries. *Electrochimica Acta* **2013**, *87*, 779–787. [CrossRef]
17. Maranski, K.; Andreev, Y.G.; Bruce, P.G. Synthesis of Poly(ethylene oxide) Approaching Monodispersity. *Angew. Chem. Int. Ed.* **2014**, *53*, 6411–6413. [CrossRef] [PubMed]
18. Zhu, H.; Chen, F.; Jin, L.; O'Dell, L.A.; Forsyth, M. Insight into Local Structure and Molecular Dynamics in Organic Solid-State Ionic Conductors. *ChemPhysChem* **2014**, *15*, 3720–3724. [CrossRef] [PubMed]
19. Mizuno, F.; Belieres, J.P.; Kuwata, N.; Pradel, A.; Ribes, M.; Angell, C.A. Highly decoupled ionic and protonic solid electrolyte systems, in relation to other relaxing systems and their energy landscapes. *J. Non-Cryst. Solids* **2006**, *352*, 5147–5155. [CrossRef]
20. Wang, Y.; Agapov, A.L.; Fan, F.; Hong, K.; Yu, X.; Mays, J.; Sokolov, A.P. Decoupling of Ionic Transport from Segmental Relaxation in Polymer Electrolytes. *Phys. Rev. Lett.* **2012**, *108*, 088303. [CrossRef] [PubMed]
21. Croce, F.; Appetecchi, G.B.; Persi, L.; Scrosati, B. Nanocomposite polymer electrolytes for lithium batteries. *Nature* **1998**, *394*, 456–458. [CrossRef]

22. Hovington, P.; Lagacé, M.; Guerfi, A.; Bouchard, P.; Mauger, A.; Julien, C.M.; Armand, M.; Zaghib, K. New Lithium Metal Polymer Solid State Battery for an Ultrahigh Energy: Nano C-LiFePO$_4$ versus Nano Li1.2V3O8. *Nano Lett.* **2015**, *15*, 2671–2678. [CrossRef] [PubMed]
23. Lago, N.; Garcia-Calvo, O.; Lopez del Amo, J.M.; Rojo, T.; Armand, M. All-Solid-State Lithium-Ion Batteries with Grafted Ceramic Nanoparticles Dispersed in Solid Polymer Electrolytes. *ChemSusChem* **2015**, *8*, 3039–3043. [CrossRef] [PubMed]
24. Bouchet, R.; Maria, S.; Meziane, R.; Aboulaich, A.; Lienafa, L.; Bonnet, J.-P.; Phan, T.N.T.; Bertin, D.; Gigmes, D.; Devaux, D.; et al. Single-ion BAB triblock copolymers as highly efficient electrolytes for lithium-metal batteries. *Nat. Mater.* **2013**, *12*, 452–457. [CrossRef] [PubMed]
25. Javier, A.E.; Patel, S.N.; Hallinan, D.T.; Srinivasan, V.; Balsara, N.P. Simultaneous Electronic and Ionic Conduction in a Block Copolymer: Application in Lithium Battery Electrodes. *Angew. Chem. Int. Ed.* **2011**, *50*, 9848–9851. [CrossRef] [PubMed]
26. Sakuda, J.; Hosono, E.; Yoshio, M.; Ichikawa, T.; Matsumoto, T.; Ohno, H.; Zhou, H.; Kato, T. Liquid Crystals: Liquid-Crystalline Electrolytes for Lithium-Ion Batteries: Ordered Assemblies of a Mesogen-Containing Carbonate and a Lithium Salt (Adv. Funct. Mater. 8/2015). *Adv. Funct. Mater.* **2015**, *25*, 1205. [CrossRef]
27. Kobayashi, Y.; Seki, S.; Mita, Y.; Ohno, Y.; Miyashiro, H.; Charest, P.; Guerfi, A.; Zaghib, K. High reversible capacities of graphite and SiO/graphite with solvent-free solid polymer electrolyte for lithium-ion batteries. *J. Power Sources* **2008**, *185*, 542–548. [CrossRef]
28. Seki, S.; Kobayashi, Y.; Miyashiro, H.; Mita, Y.; Iwahori, T. Fabrication of High-Voltage, High-Capacity All-Solid-State Lithium Polymer Secondary Batteries by Application of the Polymer Electrolyte/Inorganic Electrolyte Composite Concept. *Chem. Mater.* **2005**, *17*, 2041–2045. [CrossRef]
29. He, R.; Echeverri, M.; Ward, D.; Zhu, Y.; Kyu, T. Highly conductive solvent-free polymer electrolyte membrane for lithium-ion batteries: Effect of prepolymer molecular weight. *J. Membr. Sci.* **2016**, *498*, 208–217. [CrossRef]
30. Porcarelli, L.; Gerbaldi, C.; Bella, F.; Nair, J.R. Super Soft All-Ethylene Oxide Polymer Electrolyte for Safe All-Solid Lithium Batteries. *Sci. Rep.* **2016**, *6*, 19892. [CrossRef] [PubMed]
31. Porcarelli, L.; Aboudzadeh, M.A.; Rubatat, L.; Nair, J.R.; Shaplov, A.S.; Gerbaldi, C.; Mecerreyes, D. Single-ion triblock copolymer electrolytes based on poly(ethylene oxide) and methacrylic sulfonamide blocks for lithium metal batteries. *J. Power Sources* **2017**, *364*, 191–199. [CrossRef]
32. He, W.; Cui, Z.; Liu, X.; Cui, Y.; Chai, J.; Zhou, X.; Liu, Z.; Cui, G. Carbonate-linked poly(ethylene oxide) polymer electrolytes towards high performance solid state lithium batteries. *Electrochim. Acta* **2017**, *225*, 151–159. [CrossRef]
33. Wong, S.; Vaia, R.A.; Giannelis, E.P.; Zax, D.B. Dynamics in a poly(ethylene oxide)-based nanocomposite polymer electrolyte probed by solid state NMR. *Solid State Ionics* **1996**, *86–88*, 547–557. [CrossRef]
34. Bloembergen, N.; Purcell, E.M.; Pound, R.V. *Relaxation Effects in Nuclear Magnetic Resonance Absorption*; World Scientific Publishing Co. Pte. Ltd.: Singapore, 1990; pp. 411–444.
35. Stejskal, E.O.; Tanner, J.E. Spin Diffusion Measurements: Spin Echoes in the Presence of a Time-Dependent Field Gradient. *J. Chem. Phys.* **1965**, *42*, 288–292. [CrossRef]
36. Böhmer, R.; Jeffrey, K.R.; Vogel, M. Solid-state Li NMR with applications to the translational dynamics in ion conductors. *Prog. Nucl. Magn. Reson. Spectrosc.* **2007**, *50*, 87–174. [CrossRef]
37. Dai, Y.; Wang, Y.; Greenbaum, S.G.; Bajue, S.A.; Golodnitsky, D.; Ardel, G.; Strauss, E.; Peled, E. Electrical, thermal and NMR investigation of composite solid electrolytes based on PEO, LiI and high surface area inorganic oxides. *Electrochim. Acta* **1998**, *43*, 1557–1561. [CrossRef]
38. Croce, F.; Bonino, F.; Panero, S.; Scrosati, B. Properties of mixed polymer and crystalline ionic conductors. *Philos. Mag. B* **1989**, *59*, 161–168. [CrossRef]
39. Capuano, F.; Croce, F.; Scrosati, B. Composite Polymer Electrolytes. *J. Electrochem. Soc.* **1991**, *138*, 1918–1922. [CrossRef]
40. Wieczorek, W.; Florjanczyk, Z.; Stevens, J.R. Composite polyether based solid electrolytes. *Electrochim. Acta* **1995**, *40*, 2251–2258. [CrossRef]
41. Przyluski, J.; Siekierski, M.; Wieczorek, W. Effective medium theory in studies of conductivity of composite polymeric electrolytes. *Electrochim. Acta* **1995**, *40*, 2101–2108. [CrossRef]
42. Golodnitsky, D.; Strauss, E.; Peled, E.; Greenbaum, S. Review—On Order and Disorder in Polymer Electrolytes. *J. Electrochem. Soc.* **2015**, *162*, A2551–A2566. [CrossRef]

43. Liu, W.; Liu, N.; Sun, J.; Hsu, P.-C.; Li, Y.; Lee, H.-W.; Cui, Y. Ionic Conductivity Enhancement of Polymer Electrolytes with Ceramic Nanowire Fillers. *Nano Lett.* **2015**, *15*, 2740–2745. [CrossRef] [PubMed]
44. Liu, W.; Lee, S.W.; Lin, D.; Shi, F.; Wang, S.; Sendek, A.D.; Cui, Y. Enhancing ionic conductivity in composite polymer electrolytes with well-aligned ceramic nanowires. *Nat. Energy* **2017**, *2*, 17035. [CrossRef]
45. Weston, J.E.; Steele, B.C.H. Effects of inert fillers on the mechanical and electrochemical properties of lithium salt-poly(ethylene oxide) polymer electrolytes. *Solid State Ionics* **1982**, *7*, 75–79. [CrossRef]
46. Chen, L.; Li, Y.; Li, S.-P.; Fan, L.-Z.; Nan, C.-W.; Goodenough, J.B. PEO/garnet composite electrolytes for solid-state lithium batteries: From "ceramic-in-polymer" to "polymer-in-ceramic". *Nano Energy* **2018**, *46*, 176–184. [CrossRef]
47. Aetukuri, N.B.; Kitajima, S.; Jung, E.; Thompson, L.E.; Virwani, K.; Reich, M.-L.; Kunze, M.; Schneider, M.; Schmidbauer, W.; Wilcke, W.W.; et al. Flexible Ion-Conducting Composite Membranes for Lithium Batteries. *Adv. Energy Mater.* **2015**, *5*, 1500265. [CrossRef]
48. Zhao, Y.; Wu, C.; Peng, G.; Chen, X.; Yao, X.; Bai, Y.; Wu, F.; Chen, S.; Xu, X. A new solid polymer electrolyte incorporating $Li_{10}GeP_2S_{12}$ into a polyethylene oxide matrix for all-solid-state lithium batteries. *J. Power Sources* **2016**, *301*, 47–53. [CrossRef]
49. Tu, Z.; Kambe, Y.; Lu, Y.; Archer, L.A. Nanoporous Polymer-Ceramic Composite Electrolytes for Lithium Metal Batteries. *Adv. Energy Mater.* **2013**. [CrossRef]
50. Zheng, J.; Hu, Y.-Y. New Insights into the Compositional Dependence of Li-Ion Transport in Polymer–Ceramic Composite Electrolytes. *ACS Appl. Mater. Interfaces* **2018**, *10*, 4113–4120. [CrossRef] [PubMed]
51. Zheng, J.; Tang, M.; Hu, Y.-Y. Lithium Ion Pathway within Li7La3Zr2O12-Polyethylene Oxide Composite Electrolytes. *Angew. Chem. Int. Ed.* **2016**, *55*, 12538–12542. [CrossRef] [PubMed]
52. Villaluenga, I.; Bogle, X.; Greenbaum, S.; Gil de Muro, I.; Rojo, T.; Armand, M. Cation only conduction in new polymer–SiO_2 nanohybrids: Na^+ electrolytes. *J. Mater. Chem. A* **2013**, *1*, 8348–8352. [CrossRef]
53. Croce, F.; Sacchetti, S.; Scrosati, B. Advanced, lithium batteries based on high-performance composite polymer electrolytes. *J. Power Sources* **2006**, *162*, 685–689. [CrossRef]
54. Zhang, Y.; Lim, C.A.; Cai, W.; Rohan, R.; Xu, G.; Sun, Y.; Cheng, H. Design and synthesis of a single ion conducting block copolymer electrolyte with multifunctionality for lithium ion batteries. *RSC Adv.* **2014**, *4*, 43857–43864. [CrossRef]
55. Chen, B.; Xu, Q.; Huang, Z.; Zhao, Y.; Chen, S.; Xu, X. One-pot preparation of new copolymer electrolytes with tunable network structure for all-solid-state lithium battery. *J. Power Sources* **2016**, *331*, 322–331. [CrossRef]
56. Boden, N.; Leng, S.A.; Ward, I.M. Ionic conductivity and diffusivity in polyethylene oxode/electrolyte solutions as models for polymer electrolytes. *Solid State Ionics* **1991**, *45*, 261–270. [CrossRef]
57. Yuan, R.; Teran, A.A.; Gurevitch, I.; Mullin, S.A.; Wanakule, N.S.; Balsara, N.P. Ionic Conductivity of Low Molecular Weight Block Copolymer Electrolytes. *Macromolecules* **2013**, *46*, 914–921. [CrossRef]
58. Niitani, T.; Amaike, M.; Nakano, H.; Dokko, K.; Kanamura, K. Star-Shaped Polymer Electrolyte with Microphase Separation Structure for All-Solid-State Lithium Batteries. *J. Electrochem. Soc.* **2009**, *156*, A577–A583. [CrossRef]
59. Alloin, F.; Sanchez, J.Y.; Armand, M. Triblock copolymers and networks incorporating oligo (oxyethylene) chains. *Solid State Ionics* **1993**, *60*, 3–9. [CrossRef]
60. Daigle, J.-C.; Arnold, A.; Vijh, A.; Zaghib, K. Solid-State NMR Study of New Copolymers as Solid Polymer Electrolytes. *Magnetochemistry* **2018**, *4*, 13. [CrossRef]
61. Lin, C.-L.; Kao, H.-M.; Wu, R.-R.; Kuo, P.-L. Multinuclear Solid-State NMR, DSC, and Conductivity Studies of Solid Polymer Electrolytes Based on Polyurethane/Poly(dimethylsiloxane) Segmented Copolymers. *Macromolecules* **2002**, *35*, 3083–3096. [CrossRef]
62. Meabe, L.; Huynh, T.V.; Lago, N.; Sardon, H.; Li, C.; O'Dell, L.A.; Armand, M.; Forsyth, M.; Mecerreyes, D. Poly(ethylene oxide carbonates) solid polymer electrolytes for lithium batteries. *Electrochim. Acta* **2018**, *264*, 367–375. [CrossRef]
63. Morioka, T.; Nakano, K.; Tominaga, Y. Ion-Conductive Properties of a Polymer Electrolyte Based on Ethylene Carbonate/Ethylene Oxide Random Copolymer. *Macromol. Rapid Commun.* **2017**, *38*, 1600652. [CrossRef] [PubMed]
64. Mindemark, J.; Sun, B.; Törmä, E.; Brandell, D. High-performance solid polymer electrolytes for lithium batteries operational at ambient temperature. *J. Power Sources* **2015**, *298*, 166–170. [CrossRef]

65. Timachova, K.; Villaluenga, I.; Cirrincione, L.; Gobet, M.; Bhattacharya, R.; Jiang, X.; Newman, J.; Madsen, L.A.; Greenbaum, S.G.; Balsara, N.P. Anisotropic Ion Diffusion and Electrochemically Driven Transport in Nanostructured Block Copolymer Electrolytes. *J. Phys. Chem. B* **2018**, *122*, 1537–1544. [CrossRef] [PubMed]
66. Vashishta, P.; Mundy, J.N. *Fast Ion Transport in Solids: Electrodes and Electrolytes*; Elsevier North Holland, Inc.: Amsterdam, NY, USA, 1979.
67. Papke, B.L.; Dupon, R.; Ratner, M.A.; Shriver, D.F. Ion-pairing in polyether solid electrolytes and its influence on ion transport. *Solid State Ionics* **1981**, *5*, 685–688. [CrossRef]
68. Chintapalli, M.; Le, T.N.P.; Venkatesan, N.R.; Mackay, N.G.; Rojas, A.A.; Thelen, J.L.; Chen, X.C.; Devaux, D.; Balsara, N.P. Structure and Ionic Conductivity of Polystyrene-block-poly(ethylene oxide) Electrolytes in the High Salt Concentration Limit. *Macromolecules* **2016**, *49*, 1770–1780. [CrossRef]
69. Singh, M.; Odusanya, O.; Wilmes, G.M.; Eitouni, H.B.; Gomez, E.D.; Patel, A.J.; Chen, V.L.; Park, M.J.; Fragouli, P.; Iatrou, H.; et al. Effect of Molecular Weight on the Mechanical and Electrical Properties of Block Copolymer Electrolytes. *Macromolecules* **2007**, *40*, 4578–4585. [CrossRef]
70. Liu, T.-M.; Saikia, D.; Ho, S.-Y.; Chen, M.-C.; Kao, H.-M. High ion-conducting solid polymer electrolytes based on blending hybrids derived from monoamine and diamine polyethers for lithium solid-state batteries. *RSC Adv.* **2017**, *7*, 20373–20383. [CrossRef]
71. Nunes-Pereira, J.; Costa, C.M.; Lanceros-Méndez, S. Polymer composites and blends for battery separators: State of the art, challenges and future trends. *J. Power Sources* **2015**, *281*, 378–398. [CrossRef]
72. Nguyen, C.A.; Xiong, S.; Ma, J.; Lu, X.; Lee, P.S. High ionic conductivity P(VDF-TrFE)/PEO blended polymer electrolytes for solid electrochromic devices. *Phys. Chem. Chem. Phys.* **2011**, *13*, 13319–13326. [CrossRef] [PubMed]
73. Noor, S.A.M.; Ahmad, A.; Talib, I.A.; Rahman, M.Y.A. Morphology, chemical interaction, and conductivity of a PEO-ENR50 based on solid polymer electrolyte. *Ionics* **2010**, *16*, 161–170. [CrossRef]
74. Borodin, O.; Suo, L.; Gobet, M.; Ren, X.; Wang, F.; Faraone, A.; Peng, J.; Olguin, M.; Schroeder, M.; Ding, M.S.; et al. Liquid Structure with Nano-Heterogeneity Promotes Cationic Transport in Concentrated Electrolytes. *ACS Nano* **2017**, *11*, 10462–10471. [CrossRef] [PubMed]
75. de Souza, P.H.; Bianchi, R.F.; Dahmouche, K.; Judeinstein, P.; Faria, R.M.; Bonagamba, T.J. Solid-State NMR, Ionic Conductivity, and Thermal Studies of Lithium-doped Siloxane–Poly(propylene glycol) Organic–Inorganic Nanocomposites. *Chem. Mater.* **2001**, *13*, 3685–3692. [CrossRef]
76. Gadjourova, Z.; Andreev, Y.G.; Tunstall, D.P.; Bruce, P.G. Ionic conductivity in crystalline polymer electrolytes. *Nature* **2001**, *412*, 520–523. [CrossRef] [PubMed]
77. Christie, A.M.; Lilley, S.J.; Staunton, E.; Andreev, Y.G.; Bruce, P.G. Increasing the conductivity of crystalline polymer electrolytes. *Nature* **2005**, *433*, 50–53. [CrossRef] [PubMed]
78. Zhang, C.; Staunton, E.; Andreev, Y.G.; Bruce, P.G. Raising the Conductivity of Crystalline Polymer Electrolytes by Aliovalent Doping. *J. Am. Chem. Soc.* **2005**, *127*, 18305–18308. [CrossRef] [PubMed]
79. Staunton, E.; Andreev, Y.G.; Bruce, P.G. Factors influencing the conductivity of crystalline polymer electrolytes. *Faraday Discuss.* **2007**, *134*, 143–156. [CrossRef] [PubMed]
80. Golodnitsky, D.; Livshits, E.; Peled, E. Highly conductive oriented PEO-based polymer electrolytes. *Macromol. Symp.* **2003**, *203*, 27–46. [CrossRef]
81. Golodnitsky, D.; Livshits, E.; Ulus, A.; Barkay, Z.; Lapides, I.; Peled, E.; Chung, S.H.; Greenbaum, S. Fast Ion Transport Phenomena in Oriented Semicrystalline LiI–P(EO)n-Based Polymer Electrolytes. *J. Phys. Chem. A* **2001**, *105*, 10098–10106. [CrossRef]
82. Yan, X.; Peng, B.; Hu, B.; Chen, Q. PEO-urea-LiTFSI ternary complex as solid polymer electrolytes. *Polymer* **2016**, *99*, 44–48. [CrossRef]
83. Liu, Y.; Antaya, H.; Pellerin, C. Characterization of the stable and metastable poly(ethylene oxide)–urea complexes in electrospun fibers. *J. Polym. Sci. Part B Polym. Phys.* **2008**, *46*, 1903–1913. [CrossRef]
84. Chenite, A.; Brisse, F. Structure and conformation of poly(ethylene oxide), PEO, in the trigonal form of the PEO-urea complex at 173 K. *Macromolecules* **1991**, *24*, 2221–2225. [CrossRef]
85. Liu, Y.; Pellerin, C. Highly Oriented Electrospun Fibers of Self-Assembled Inclusion Complexes of Poly(ethylene oxide) and Urea. *Macromolecules* **2006**, *39*, 8886–8888. [CrossRef]
86. Liang, H.; Li, H.; Wang, Z.; Wu, F.; Chen, L.; Huang, X. New Binary Room-Temperature Molten Salt Electrolyte Based on Urea and LiTFSI. *J. Phys. Chem. B* **2001**, *105*, 9966–9969. [CrossRef]

87. Fu, X.-B.; Yang, G.; Wu, J.-Z.; Wang, J.-C.; Chen, Q.; Yao, Y.-F. Fast Lithium-Ion Transportation in Crystalline Polymer Electrolytes. *ChemPhysChem* **2018**, *19*, 45–50. [CrossRef] [PubMed]
88. Tong, Y.; Chen, L.; He, X.; Chen, Y. Sequential effect and enhanced conductivity of star-shaped diblock liquid-crystalline copolymers for solid electrolytes. *J. Power Sources* **2014**, *247*, 786–793. [CrossRef]
89. MacFarlane, D.R.; Forsyth, M.; Howlett, P.C.; Pringle, J.M.; Sun, J.; Annat, G.; Neil, W.; Izgorodina, E.I. Ionic Liquids in Electrochemical Devices and Processes: Managing Interfacial Electrochemistry. *Acc. Chem. Res.* **2007**, *40*, 1165–1173. [CrossRef] [PubMed]
90. MacFarlane, D.R.; Huang, J.; Forsyth, M. Lithium-doped plastic crystal electrolytes exhibiting fast ion conduction for secondary batteries. *Nature* **1999**, *402*, 792–794. [CrossRef]
91. Chen, F.; Pringle, J.M.; Forsyth, M. Insights into the Transport of Alkali Metal Ions Doped into a Plastic Crystal Electrolyte. *Chem. Mater.* **2015**, *27*, 2666–2672. [CrossRef]
92. Wanakule, N.S.; Panday, A.; Mullin, S.A.; Gann, E.; Hexemer, A.; Balsara, N.P. Ionic Conductivity of Block Copolymer Electrolytes in the Vicinity of Order−Disorder and Order−Order Transitions. *Macromolecules* **2009**, *42*, 5642–5651. [CrossRef]
93. Choi, I.; Ahn, H.; Park, M.J. Enhanced Performance in Lithium–Polymer Batteries Using Surface-Functionalized Si Nanoparticle Anodes and Self-Assembled Block Copolymer Electrolytes. *Macromolecules* **2011**, *44*, 7327–7334. [CrossRef]
94. Sun, J.; Liao, X.; Minor, A.M.; Balsara, N.P.; Zuckermann, R.N. Morphology-Conductivity Relationship in Crystalline and Amorphous Sequence-Defined Peptoid Block Copolymer Electrolytes. *J. Am. Chem. Soc.* **2014**, *136*, 14990–14997. [CrossRef] [PubMed]
95. Tong, Y.; Chen, L.; He, X.; Chen, Y. Free Mesogen Assisted Assembly of the Star-shaped Liquid-crystalline Copolymer/Polyethylene Oxide Solid Electrolytes for Lithium Ion Batteries. *Electrochim. Acta* **2014**, *118*, 33–40. [CrossRef]
96. Lin, D.; Liu, W.; Liu, Y.; Lee, H.R.; Hsu, P.-C.; Liu, K.; Cui, Y. High Ionic Conductivity of Composite Solid Polymer Electrolyte via In Situ Synthesis of Monodispersed SiO_2 Nanospheres in Poly(ethylene oxide). *Nano Lett.* **2016**, *16*, 459–465. [CrossRef] [PubMed]
97. MacGlashan, G.S.; Andreev, Y.G.; Bruce, P.G. Structure of the polymer electrolyte poly(ethylene oxide)$_6$:LiAsF$_6$. *Nature* **1999**, *398*, 792–794. [CrossRef]
98. Yang, L.-Y.; Wei, D.-X.; Xu, M.; Yao, Y.-F.; Chen, Q. Transferring Lithium Ions in Nanochannels: A PEO/Li$^+$ Solid Polymer Electrolyte Design. *Angew. Chem.* **2014**, *126*, 3705–3709. [CrossRef]
99. Fu, X.-B.; Yang, L.-Y.; Ma, J.-Q.; Yang, G.; Yao, Y.-F.; Chen, Q. Revealing structure and dynamics in host–guest supramolecular crystalline polymer electrolytes by solid-state NMR: Applications to β-CD-polyether/Li$^+$ crystal. *Polymer* **2016**, *105*, 310–317. [CrossRef]
100. Okumura, H.; Kawaguchi, Y.; Harada, A. Preparation and Characterization of Inclusion Complexes of Poly(dimethylsiloxane)s with Cyclodextrins. *Macromolecules* **2001**, *34*, 6338–6343. [CrossRef]
101. Li, J.; Ni, X.; Zhou, Z.; Leong, K.W. Preparation and Characterization of Polypseudorotaxanes Based on Block-Selected Inclusion Complexation between Poly(propylene oxide)-Poly(ethylene oxide)-Poly(propylene oxide) Triblock Copolymers and α-Cyclodextrin. *J. Am. Chem. Soc.* **2003**, *125*, 1788–1795. [CrossRef] [PubMed]
102. Chen, L.; Zhu, X.; Yan, D.; Chen, Y.; Chen, Q.; Yao, Y. Controlling Polymer Architecture through Host–Guest Interactions. *Angew. Chem. Int. Ed.* **2006**, *45*, 87–90. [CrossRef] [PubMed]
103. Murugan, R.; Thangadurai, V.; Weppner, W. Fast Lithium Ion Conduction in Garnet-Type $Li_7La_3Zr_2O_{12}$. *Angew. Chem. Int. Ed.* **2007**, *46*, 7778–7781. [CrossRef] [PubMed]
104. Zhang, C.; Gamble, S.; Ainsworth, D.; Slawin, A.M.Z.; Andreev, Y.G.; Bruce, P.G. Alkali metal crystalline polymer electrolytes. *Nat. Mater.* **2009**, *8*, 580–584. [CrossRef] [PubMed]
105. Pope, C.R.; Romanenko, K.; MacFarlane, D.R.; Forsyth, M.; O'Dell, L.A. Sodium ion dynamics in a sulfonate based ionomer system studied by ^{23}Na solid-state nuclear magnetic resonance and impedance spectroscopy. *Electrochim. Acta* **2015**, *175*, 62–67. [CrossRef]
106. Tiyapiboonchaiya, C.; Pringle, J.M.; MacFarlane, D.R.; Forsyth, M.; Sun, J. Polyelectrolyte-in-Ionic-Liquid Electrolytes. *Macromol. Chem. Phys.* **2003**, *204*, 2147–2154. [CrossRef]
107. Ünal, H.I.; Yilmaz, H. Electrorheological properties of poly(lithium-2-acrylamido-2-methyl propane sulfonic acid) suspensions. *J. Appl. Polym. Sci.* **2002**, *86*, 1106–1112. [CrossRef]

108. Noor, S.A.M.; Sun, J.; MacFarlane, D.R.; Armand, M.; Gunzelmann, D.; Forsyth, M. Decoupled ion conduction in poly(2-acrylamido-2-methyl-1-propane-sulfonic acid) homopolymers. *J. Mater. Chem. A* **2014**, *2*, 17934–17943. [CrossRef]
109. Wang, W.; Tudryn, G.J.; Colby, R.H.; Winey, K.I. Thermally Driven Ionic Aggregation in Poly(ethylene oxide)-Based Sulfonate Ionomers. *J. Am. Chem. Soc.* **2011**, *133*, 10826–10831. [CrossRef] [PubMed]
110. Tudryn, G.J.; Liu, W.; Wang, S.-W.; Colby, R.H. Counterion Dynamics in Polyester–Sulfonate Ionomers with Ionic Liquid Counterions. *Macromolecules* **2011**, *44*, 3572–3582. [CrossRef]
111. Mandai, T.; Nozawa, R.; Tsuzuki, S.; Yoshida, K.; Ueno, K.; Dokko, K.; Watanabe, M. Phase Diagrams and Solvate Structures of Binary Mixtures of Glymes and Na Salts. *J. Phys. Chem. B* **2013**, *117*, 15072–15085. [CrossRef] [PubMed]
112. Freitag, K.M.; Walke, P.; Nilges, T.; Kirchhain, H.; Spranger, R.J.; van Wüllen, L. Electrospun-sodiumtetrafluoroborate-polyethylene oxide membranes for solvent-free sodium ion transport in solid state sodium ion batteries. *J. Power Sources* **2018**, *378*, 610–617. [CrossRef]
113. Freitag, K.M.; Kirchhain, H.; Wüllen, L.V.; Nilges, T. Enhancement of Li Ion Conductivity by Electrospun Polymer Fibers and Direct Fabrication of Solvent-Free Separator Membranes for Li Ion Batteries. *Inorg. Chem.* **2017**, *56*, 2100–2107. [CrossRef] [PubMed]

© 2018 by the authors. Licensee MDPI, Basel, Switzerland. This article is an open access article distributed under the terms and conditions of the Creative Commons Attribution (CC BY) license (http://creativecommons.org/licenses/by/4.0/).

Article

High Performance Polymer/Ionic Liquid Thermoplastic Solid Electrolyte Prepared by Solvent Free Processing for Solid State Lithium Metal Batteries

Francisco González [1], Pilar Tiemblo [1,*], Nuria García [1], Oihane Garcia-Calvo [2], Elisabetta Fedeli [2], Andriy Kvasha [2,*] and Idoia Urdampilleta [2]

1. Instituto de Ciencia y Tecnología de Polímeros, ICTP-CSIC, Juan de la Cierva 3, 28006 Madrid, Spain; fgonzalez@ictp.csic.es (F.G.); ngarcia@ictp.csic.es (N.G.)
2. CIDETEC Energy Storage, Parque Científico y Tecnológico de Gipuzkoa, Paseo Miramón 196, 20009 Donostia-San Sebastián, Spain; ogarcia@cidetec.es (O.G.-C.); efedeli@cidetec.es (E.F.); iurdampilleta@cidetec.es (I.U.)
* Correspondence: ptiemblo@ictp.csic.es (P.T.); akvasha@cidetec.es (A.K.); Tel.: +34-915-618-806 (P.T.); +34-943-309-022 (A.K.)

Received: 6 July 2018; Accepted: 26 July 2018; Published: 2 August 2018

Abstract: A polymer/ionic liquid thermoplastic solid electrolyte based on poly(ethylene oxide) (PEO), modified sepiolite (TPGS-S), lithium bis(trifluoromethanesulfonyl)imide (LiTFSI), and 1-Butyl-1-methylpyrrolidinium bis(trifluoromethanesulfonyl)imide ($PYR_{14}TFSI$) ionic liquid is prepared using solvent free extrusion method. Its physical-chemical, electrical, and electrochemical properties are comprehensively studied. The investigated solid electrolyte demonstrates high ionic conductivity together with excellent compatibility with lithium metal electrode. Finally, truly Li-LiFePO$_4$ solid state coin cell with the developed thermoplastic solid electrolyte demonstrates promising electrochemical performance during cycling under 0.2 C/0.5 C protocol at 60 °C.

Keywords: solid state battery; thermoplastic polymer electrolyte; ionic liquid; sepiolite; inorganic filler

1. Introduction

Highly efficient, light, safe, and long-lasting rechargeable batteries are the goal of all the researchers and producers involved in the energy storage business. So far, lithium ion batteries (LIBs) represent the most promising answer; however, the booming growth of demand spotlighted the drawbacks of such technology. The major intrinsic limitation of LIBs is the low theoretical specific capacity (372 mAh·g^{-1}) of the traditional graphite anode, which does not allow the increase of practical LIB energy density to more than 300 Wh·kg^{-1}. Lithium metal represents the best alternative anode material to produce high energy density batteries because it possesses the lowest standard potential (E_o = −3.04 V versus standard hydrogen electrode) and the highest theoretical capacity (3.860 mAh^{-1}) [1]. Unfortunately, this technology is not ideal and presents several issues such as dendrite growth, instability of lithium metal with the most part of classical organic liquid electrolytes, low coulombic efficiency, poor cyclability, and poor safety due to leakage and high flammability of the liquid electrolyte based on a mixture of carbonate solvents [2–4].

Solid polymer electrolytes (SPEs) are without a doubt among the key solutions to overcome such limitations toward high energy density, efficient, and safe solid state batteries (SSB) [4–7]. Indeed, these solid ion conductive membranes can replace microporous separators impregnated by volatile flammable organic electrolytes [5,6], acting as physical barrier against dendrite growth reducing the possibility of short-circuit, thermal runaway, and explosion, significantly improving the safety of the battery [8–10]. However, poor ionic conductivity at room temperature due to low mobility of the lithium cations in the solid polymer matrix, and the loss of mechanical properties in the conductive molten state at higher temperature, limit their spread in the battery market [11]. Several solutions have been proposed to increase ionic conductivity while maintaining good mechanical properties [12,13]. Many of them are based on the addition of low molecular weight compounds with adequate electrochemical properties coupled to the creation of physical or chemical crosslinking sites at the polymer [14]. The employment of inorganic fillers and the introduction of sufficient amount of low molecular weight compounds are listed among the most relevant examples. Adding inorganic fillers proved to favor the performance of the battery by (i) preventing crystallization by hindering the supramolecular arrangement of the polymer chains; (ii) favoring ionic dissociation, improving the matrix/solid electrolyte interface (SEI) interaction thanks to the contribution of different possible surface groups; and (iii) increasing mechanical resistance and stability [15–17]. The employment of low molecular weight compounds also proved to be an effective measure to enhance the electrochemical performance of a solid state battery [18]. Among them, and despite some drawbacks (high cost and some instability at lithium deposition potential [19]), room temperature ionic liquids (RTILs) are, probably, the most promising materials thanks to their negligible vapor pressure, low flammability, high ionic conductivity in comparison with solid polymer electrolytes, and their ability to form an effective solid electrolyte interphase onto the lithium metal electrode surface [20–22]. Several recent studies demonstrated that the presence of RTILs can enhance significantly the electrochemical properties of the solid state battery, such as, for instance, improving the long-time stability in the stripping/plating from lithium metal electrodes [23,24]. Furthermore, it has recently been demonstrated that free RTILs in a polymeric solid matrix can undergo percolation, creating a highly conductive pathway across the electrolyte and a wet interfacial layer that greatly improves the interfacial compatibility with the electrodes [25].

On the other hand, the always increasing demand of electronic devices goes together with a growing concern about a sustainable future and cost considerations. The combination of these two factors bursts the research toward the development of green processes to obtain high performance materials. In this optic, fast production methods that employ recyclable materials and reduce or eliminate completely the use of harmful organic solvents are the main goal of all the efforts. Thermoplastic polymers represent a valid option to develop solid electrolytes because they can be processed easily by extrusion and shaped by hot-pressing or lamination, none of which require solvent, and can be theoretically recycled and reprocessed, in this way reducing the final cost of solid state batteries, which is crucial for their implementation in the market [26–28].

In this context, this work presents the solvent-free preparation of a thermoplastic polymer electrolyte (TPE) consisting on a polymeric matrix, ad hoc modified inorganic fillers, an ionic liquid, and a lithium salt. More precisely, the TPE is composed by poly(ethylene oxide) (PEO), surface modified sepiolite (TPGS-S), lithium bis(trifluoromethanesulfonyl)imide (LiTFSI), and 1-Butyl-1-methylpyrrolidinium bis(trifluoromethanesulfonyl)imide ($PYR_{14}TFSI$), prepared by solvent-free extrusion method. This TPE is compared with a well-studied reference electrolyte consisting of PEO and LiTFSI [29]. The extensive physical and electrochemical characterization of the new TPE is presented in this article. The developed solid electrolyte demonstrated high ionic conductivity, good electrochemical stability, excellent compatibility with lithium metal, and promising cycling performance in truly solid state Li-$LiFePO_4$ coin cell prototype.

2. Materials and Methods

2.1. Reagents

For the preparation of the solid electrolytes, the following materials were used: PEO: Mn 5×10^6 g·mol^{-1} for the TPE, Mn 6×10^5 g·mol^{-1} for the reference electrolyte, and Mn 4×10^5 g·mol^{-1}

for the positive electrode preparation, all purchased from Sigma-Aldrich (St. Louis, MO, USA). D-α-tocopherol polyethylene glycol 1000 succinate (TPGS), used to prepare the modified sepiolite (TPGS-S), was purchased from Sigma-Aldrich and used as received. Details on the preparation of the TPGS-S have appeared elsewhere [30]. Battery grade LiTFSI and PYR$_{14}$TFSI with 99.9% of purity were purchased from Solvionic (Toulouse, France). Dry acetonitrile with 99.8% of purity was purchased from Scharlab (Barcelona, Spain). All the reagents were stored in dry room with dew point below −50 °C; they were used without further purification.

2.2. Synthesis and Preparation of Materials

Reference solid polymer electrolyte (PEO-LiTFSI) was prepared as follows: LiTFSI was dissolved in acetonitrile and stirred with a mechanical stirrer for 30 min. PEO, Mn 6×10^5 g·mol^{-1}, was slowly added and the mixture was stirred for 5 h to guarantee the complete solubilization of all reagents. The molar ratio of EO/Li was chosen to be 20. The amount of solid in the acetonitrile solution was set to 12 wt %. Self-standing membranes of reference PEO-LiTFSI electrolyte were obtained by solvent casting over Teflon sheets. The casted solution was dried for 2 h at 35 °C and then for 17 h at 60 °C under reduced pressure. PEO-LiTFSI electrolyte formulation is given in Table 1.

TPE was prepared in accordance with method reported earlier [26]. Briefly, all components were physically premixed and then melt compounded in a Haake MiniLab extruder (Haake Minilab, Thermo Fisher Scientific, Waltham, MA, USA). Processing was carried out at a shear rate of 80 rpm during 20 min and at 160 °C. Afterwards, TPE extrudate was processed by hot pressing at 75 °C. TPE electrolyte formulation is given in Table 1.

Table 1. Main features of the investigated solid electrolytes. PYR14TFSI—1-Butyl-1-methylpyrrolidinium bis(trifluoromethanesulfonyl)imide; LiTFSI—lithium bis(trifluoromethanesulfonyl) imide; PEO—poly (ethylene oxide); TPGS-S—surface modified sepiolite; DSC—differential scanning calorimetry; TPE—thermoplastic polymer electrolyte.

Solid Electrolyte	PYR14TFSI mol m^{-3}	LiTFSI mol m^{-3}	PEO mol m^{-3}	TPGS-S wt %	DSC Xc/Tm (%/°C)	σ (25 °C) mS cm^{-1}	σ (60 °C) mS cm^{-1}
PEO-LiTFSI	0	892	20670	0	32/53	0.01	0.5
TPE	1577	790	9826	2.5	5/38	0.50	3.0

2.3. Physicochemical Characterization

Characterization of electrolytes was done on films of controlled thickness processed by compression molding at 75 °C during 3 min.

Scanning electron microscopy (SEM) was performed with a Hitachi SU-8000 (Hitachi Ltd., Tokyo, Japan). Samples were fractured after immersion in liquid nitrogen and the sections were observed unmetalized.

Differential scanning calorimetry (DSC) studies were performed in a TA Instruments Q100 (TA Instruments, New Castle, DE, USA). The heat flow was recorded as follows: two cooling-heating cycles at 10 °C·min^{-1} from 120 °C to −80 °C, followed by a second cooling-heating cycle from 120 °C to −80 °C at 20 °C·min^{-1}. DSC data included in Table 1 were obtained from the second DSC heating trace at 10 °C·min^{-1}. The crystallinity percentage (χ_c) was determined considering 100% crystalline PEO heat of melting as $\Delta H_m = 197$ J·g^{-1} [31]. The % χ_c in Table 1 is referred to the weight of the electrolyte and not to the weight fraction of PEO.

Thermogravimetric analysis (TGA) was performed in a TA Q-500 in nitrogen atmosphere at 10 °C·min^{-1} up to 800 °C.

Determination of diffusion coefficients (D) was done by ^{7}Li and ^{19}F pulsed field gradient-NMR (PFG-NMR) in a Bruker AvanceTM 400 spectrometer (Bruker BioSpin GmbH, Rheinstetten, Germany) as reported before [26]. The lithium transference number measured by NMR ($t_{Li^+}^{NMR}$) was calculated using Equation (1). It was not possible to measure D of the cation (D_{Pyr}), because of the overlapping with PEO protons, so it was estimated to be about 10% lower than TFSI, according to bibliographic data [32].

$$t_{Li^+}^{NMR} = \frac{D_{Li}c_{Li}}{D_{Li}c_{Li} + D_{TFSI}c_{TFSI} + D_{Pyr}c_{Pyr}} \quad (1)$$

Creep experiments were done as follows: electrolyte films of about 500 μm were sandwiched between two gold electrodes of 20 mm of diameter, and placed on a heating plate with a 0.5 kg load on top and kept 20 min at 70 °C, followed by 20 min at 90 °C.

2.4. Electrochemical Characterization

The ionic conductivity of the TPE and PEO-LiTFSI electrolytes was determined by electrochemical impedance in a NOVOCONTROL GmbH Concept 40 broadband dielectric spectrometer (Novocontrol Technologies GmbH, Montabaur, Germany) in the temperature range of 50 °C to 90 °C and in the frequency range between 0.1 Hz and 10^7 Hz. Disk films of dimensions of 2 cm diameter and ~500 μm thickness were inserted between two gold-plated flat electrodes, then a frequency sweep was done every 10 °C, cooling to −50 °C and then heating to 90 °C; thereafter, the same measurements were done but cooling from 85 °C to 25 °C. Ionic conductivity was calculated by using conventional methods based on the Nyquist diagram and the phase angle as a function of the frequency plot. The values that appear in Table 1 correspond to the second heating measurement.

Lithium transference number (t_{Li^+}) of the TPE was obtained at 60 °C by combined alternating current (AC) impedance and direct current (DC) polarization measurements using a Solartron Analytical 1400 CellTest System (cell test, City, UK) coupled with frequency response analyzer 1455 (Ametek) of a symmetrical solid state Li/TPE/Li coin cell (2025, Hohsen, City, Japan). Coin cells were prepared using high-purity lithium metal foil (Albermale, Charlotte, NC, USA) with thickness of 50 μm. Before the measurement, the assembled coin cells were kept at 60 °C overnight to achieve a good contact and stable interface between the solid electrolyte and lithium metal electrodes. Successively, a DC potential (ΔV = 5 mV) was applied until a steady current was obtained; then, initial (I_o, after 5 milliseconds) and steady state (I_{ss}) currents that flow through the cell were measured. Impedance spectra were recorded (from 1 MHz to 1 Hz) with 10 mV sinusoidal amplitude before and after DC polarization. Subsequently, initial (R_o) and final (R_{ss}) bulk resistances of the electrolyte, and initial (R_{Co}) and final (R_{Css}) charge transfer resistances (Ω) of the interfacial layers Li metal electrode/electrolyte were derived from electrochemical impedance spectra using ZView software 3.5 (Scribner, Southern Pines, NC, USA) Using these measured values, t_{Li^+} was calculated by the following Equation (2) [33,34].

$$t_{Li^+} = \frac{I_{ss} \cdot R_{ss} \cdot (\Delta V - I_o \cdot R_{Co})}{I_o \cdot R_o \cdot (\Delta V - I_{ss} \cdot R_{Css})} \quad (2)$$

The electrochemical stability window of the TPE was evaluated in three-electrode cells using a Solartron Analytical 1400 CellTest System (Ametek) coupled with a frequency response analyzer 1455 (Ametek). To do so, a solid-state three electrode cell (HS-3E, Hohsen), using stainless steel as a working electrode, a lithium metal (50 μm) disc as a counter electrode, a lithium metal ring as a reference electrode, and a solid electrolyte membrane (80–100 μm) placed between electrodes was fabricated. The cyclic voltammetry (CV) test was carried out at a linear scan rate of 1 mV·s^{-1} to determine the electrochemical performance in cathodic (from OCV to −0.5 V) range. The oxidation stability of the investigated solid electrolyte was determined by linear sweep voltammetry (LSV) from OCV to 6 V at a scan rate of 1 mV·s^{-1}. Both CV and LSV experiments were performed at 60 °C using different TPE samples.

Galvanostatic stripping-plating studies were carried out at 60 °C in a symmetrical Li/TPE/Li coin cell (2025, Hohsen), using two lithium metal discs (Albermale, high-purity foil, 50 μm) and TPE films (80–100 μm) placed in between. The measurements were performed with the help of BaSyTec cell test system (BaSyTec, Asselfingen, Germany) at 60 °C. Galvanostatic cycles were run by applying symmetrical 1 mA·cm^{-2} current for 2 h with depth of cycling of 2 mAh·cm^{-2}.

Galvanostatic charge-discharge test in solid-state coin cells with lithium metal anode (Albermale, high-purity foil, 50 μm) and composite LiFePO$_4$ (LFP) cathode was performed at 60 °C using the BaSyTec cell test system. The cathode consisted of micro-scale carbon coated LFP material (D50: 2–4 μm), PEO-LiTFSI solid electrolyte (EO/Li~20) as ionic conductive binder, and carbon black as a conductive additive. Superficial capacity of the prepared positive electrode was 0.5 mAh·cm^{-2}. A carbon coated aluminum current collector was used to enhance interfacial resistance and avoid aluminum corrosion in the presence of TFSI anions. Solid-state coin cells were assembled in a dry room with dew point below −50 °C. Once assembled, the cells were kept for 3 h at 60 °C and then cycled within the 2.5–3.8 V range at the same temperature using BaSyTec cell test system. It is important to note that cell design, assembly, and formation procedures were not optimized in this study.

3. Results

3.1. Physicochemical Investigation

Similar TPE reported before [23,25] have shown two properties that make them interesting candidates as electrolytes, their liquid nature at the microscopic scale and their ability to remain as solids at the macroscale up to 90 °C for long periods of time [23]. Figure 1 summarizes the physicochemical characterization performed with both the TPE under study and the reference PEO-LiTFSI, which includes a SEM micrograph of the TPE, and the TGA, DSC, and creep experiments of both electrolytes.

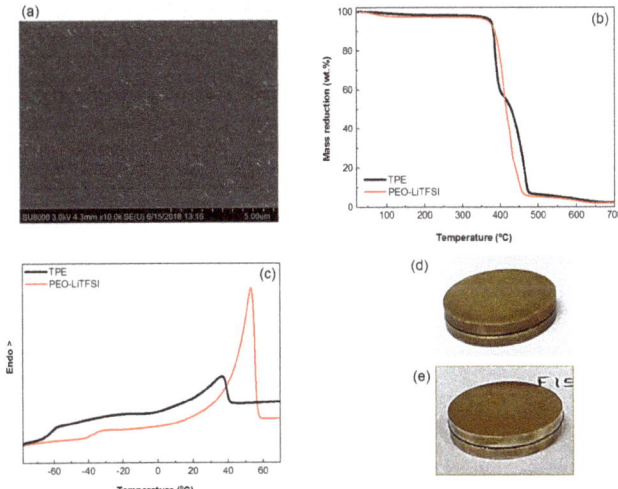

Figure 1. (a) Scanning electron microscopy (SEM) image of the thermoplastic polymer electrolyte (TPE) (cross-section); (b) thermogravimetric analysis (TGA) of the poly(ethylene oxide) (PEO)-lithium bis(trifluoromethanesulfonyl)imide (LiTFSI) and TPE; (c) differential scanning calorimetry (DSC) traces of the PEO-LiTFSI and TPE; and (d) pictures showing the electrolyte appearance after the creep test (see text for details).

First of all, Figure 1a shows the excellent dispersion of the sepiolite nanofibers in the electrolyte. Figure 1b shows that the thermal stabilities of the TPE and PEO-LiTFSI in nitrogen are very similar and will favor the overall solid-state battery safety. Figure 1c shows how the TPE has two well defined transitions, the PEO glass transition close to −60 °C and a melting endotherm slightly under 40 °C caused by the scarce crystalline phase in the TPE. On its turn, PEO-LiTFSI has a T_g at about −38 °C and a melting endotherm at 50 °C, the latter caused by the crystalline PEO phase, which amounts to 31%. Both the higher T_g and the higher crystalline fraction of PEO-LiTFSI make this electrolyte more rigid than TPE. Figure 1d and 1e show the appearance of the sandwiches (electrode-electrolyte) of PEO-LiTFSI and TPE, respectively, after the creep tests. No creep is seen in either sample after being subjected to the temperature cycles under pressure.

The ionic diffusivity in the TPE has been obtained by PFG-NMR experiments at 25 °C, and values are in the range of those obtained for similar TPE [23]: $D_{Li} = 0.6 \times 10^{-12}$ m^2·s^{-1} and $D_{TFSI} = 3.9 \times 10^{-12}$ m^2·s^{-1}. A transport number $t_{Li^+} = 0.03$ at 25 °C can be estimated from these diffusion coefficients using Equation (1).

3.2. Electrochemical Investigation

Figure 2 shows ionic conductivity (σ) values of the TPE and PEO-LiTFSI on heating from −50 °C to 90 °C. Ionic conductivity data of PYR$_{14}$TFSI [35] shown in Figure 2 demonstrates its higher conductivity comparison with both solid electrolytes. The σ of TPE increases up to values close to 10^{-2} S·cm^{-1}, and likewise decreases on going from 90 °C to 25 °C, where it attains a value of 5×10^{-4} S·cm^{-1}. As a consequence of the very low fraction of TPE suffering phase transitions in the temperature range studied, the heating and the cooling cycle measurements produce the same σ values, and so also with regards to σ variation with temperature, the TPE can be considered as a liquid. On the contrary, PEO-LiTFSI suffers the melting of the crystalline phase at about 50 °C on heating, and on cooling, an abrupt decrease of σ is seen below 50 °C, caused by the crystallization of PEO. Hence, the cooling and heating scans do not coincide in the vicinity of the phase transition. As a consequence, under 50 °C, the difference in σ between the TPE and PEO-LiTFSI becomes progressively higher.

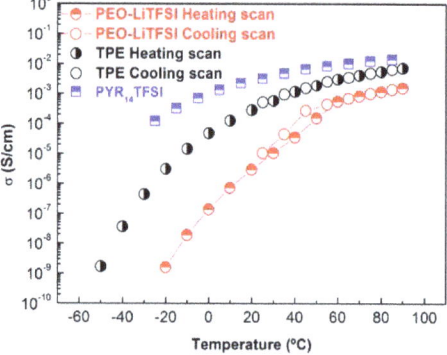

Figure 2. Ionic conductivity of the TPE and PEO-LiTFSI on heating and cooling scans. Ionic conductivity of 1-Butyl-1-methylpyrrolidinium bis(trifluoromethanesulfonyl)imide (PYR14TFSI) data reported by Martinelli et al [35] is given for comparison.

The t_{Li^+} is a very important characteristic of an electrolyte. A higher t_{Li^+} can reduce concentration polarization during charge/discharge steps and, consequently, can increase power density. Moreover, it can hinder Li metal dendrite growth and avoid decomposition and precipitation of the lithium salt. Figure 3a depicts the chronoamperometry of the symmetrical Li-Li coin cell with the investigated TPE. The AC impedance spectra before and after polarization of the cell are exhibited in Figure 3b. The equivalent circuit used for the determination of R_o, R_{ss}, R_{Co} and R_{Css} values is shown as an inset in Figure 3b.

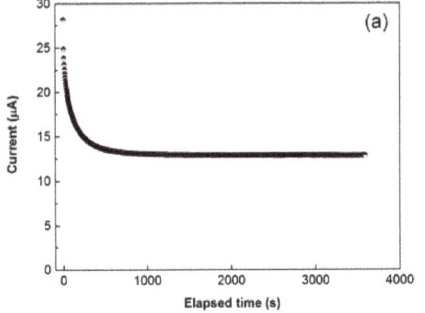 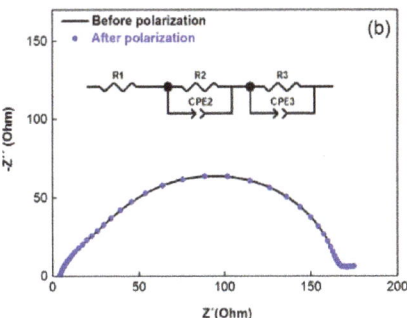

Figure 3. (a) Chronoamperometry of the Li/TPE/Li cell; (b) the alternative current (AC) impedance spectra before and after polarization. Inset (b): the equivalent circuit used for the fitting of the spectra, R_1 corresponds to R_0, R_{ss} is the sum of R_2 and R_3 to R_{Co} and R_{Css}. Test was performed at 60 °C.

Lithium ion transference number of the investigated TPE with EO/Li ratio 12, as obtained from Equation (2), is 0.08 ± 0.01 at 60 °C, which is lower than the value obtained for PEO-LiTFSI (EO/Li ~20), which is 0.25 ± 0.01. Such a difference is the direct consequence of the lower molar fraction of Li cations in the TPE (χ_{Li^+} = 0.17) than in PEO-LiTFSI (χ_{Li^+} = 0.5) or, in other words, while in the reference electrolyte, 50% of the ionic species are Li cations, the amount of Li cations in the TPE is only the 17% of the ionic species. A relatively low transport number (t_{Li^+}) of TPE does not mean lower lithium ion conductivity (σ_{Li^+}) in comparison with PEO-LiTFSI. In particular, as can be seen in Table 2, TPE has 1.9 times higher Li-ion conductivity compared with the reference at 60 °C. Moreover, we anticipate that the difference will increase over an order of magnitude at temperature below 60 °C, when the reference electrolyte will be partly crystalline because of PEO, while the TPE will remain amorphous.

Table 2. Ionic conductivity values of the investigated solid electrolytes.

Solid Electrolyte	σ (60 °C) mS cm^{-1}	σ_{Li^+} (60 °C) [a] mS cm^{-1}
PEO-LiTFSI	0.5	0.125
TPE	3.0	0.240

Note: [a] t^+ calculated by combined alternative current (AC) impedance and direct current (DC) polarization measurements reported above.

The electrochemical stability of the electrolyte is a fundamental property that determines the electrochemical behavior of the whole solid state battery. Figure 4 shows the electrochemical stability of the TPE under study towards anodic oxidation and cathodic reduction reactions. From the cathodic profile, reversible lithium plating and stripping processes are well evidenced. On the other hand, anodic LSV scab showed that the investigated electrolyte is anodically stable up to 4.2 V, which is a typical value for PEO based solid electrolytes.

Lithium metal electrode at contact with unappropriated solid electrolyte may show quite poor electrochemical behaviour (low coulombic efficiency, poor reversibility, and even lithium dendrite growth) due to cycling conditions (temperature, current density, depth of cycling) and properties of solid electrolyte layer (SEI) formed at lithium/solid electrolyte interface (nature, thickness, resistance etc.). Therefore, in this work, the compatibility of the TPE with the Li metal anode was evaluated by a galvanostatic stripping-plating test performed in Li-Li symmetrical coin cell with cycling conditions (temperature, current density, and depth of cycling) similar with full solid state cell application. Striping-plating curves for several separated cycles shown in Figure 5 with the aim of highlighting their evolution during the test. The TPE demonstrated quite low polarization and long term cyclability during more than 400 cycles (>1600 h) under relatively harsh cycling conditions.

This result demonstrates that this solid electrolyte is well compatible with lithium metal anode that is necessary requirement for its further application in lithium metal solid state batteries.

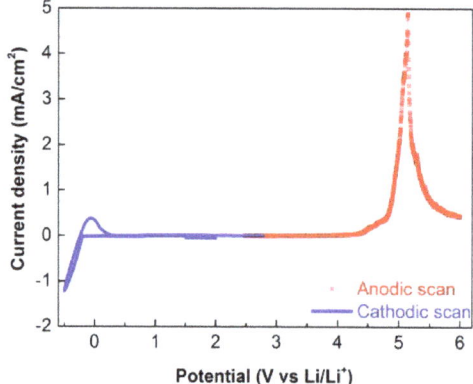

Figure 4. Cyclic voltammetry (CV) (blue) and linear sweep voltammetry (LSV) (red) curves of the TPE measured at 60 °C.

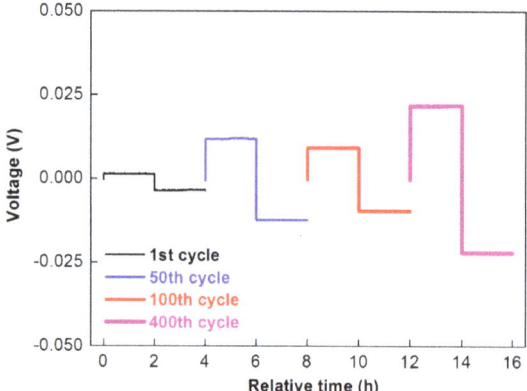

Figure 5. Cell voltage versus test time of lithium striping-plating in a symmetrical coin cell Li/TPE/Li measured on 1st, 50th, 100th, and 400th cycles. Cycling conditions: 60 °C, current density: 1 mA·cm^{-2}, duration of each step 2 h.

Finally, the TPE was tested at 60 °C in all-solid-state coin cell with lithium metal anode and composite LiFePO$_4$ cathode. Figure 6a presents charge-discharge curves of solid state coin cells with PEO-LiTFSI and TPE solid electrolytes. On the first cycle, polarization of the cell based on TPE is slightly higher in comparison with reference PEO-LiTFSI. However, during following cycles, charge-discharge profiles of both cells became quite similar.

The cycling performance of the solid state cell with TPE and PEO-LiTFSI is shown in Figure 6b. As it can be observed, upon cycling, solid state coin cell with TPE demonstrated more stable and higher coulombic efficiency (Figure 6a,c) and, as a result, much better electrochemical performance compared with the cell based on the reference PEO-LiTFSI compound.

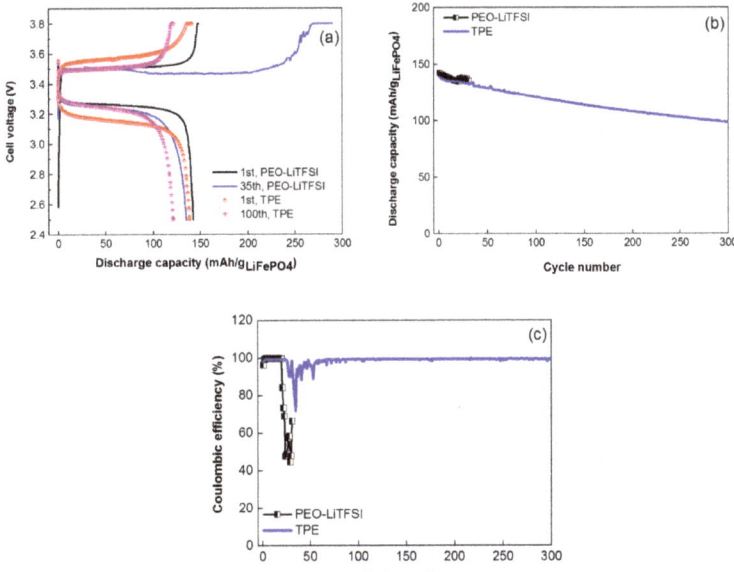

Figure 6. Electrochemical performance of solid-state coin cells Li-LiFePO$_4$: (**a**) charge-discharge curves; (**b**) specific discharge capacity; and (**c**) coulombic efficiency versus cycle number. Cycling conditions: 60 °C, Constant Current-Constant Voltage (CCCV) charge at 0.2 C (charge current cut off 0.1 C), discharge at 0.5 C, cycling interval 2.5–3.8 V, positive electrode loading 0.5 mAh·cm^{-2}.

It should be noted that relatively fast capacity decay of the nonoptimized solid-state coin cell prototype with TPE electrolyte could be related to several reasons, such as solid electrolyte impurities; traces of water in the electrolyte and cathode; and possible restructurization of TPE, which cointans noticeable amount of PYR$_{14}$TFSI ionic liquid [23]. We believe that further optimization of solid state cell prototype and assembly-formation procedures may significantly improve its electrochemical performance and durability.

Thus, our investigation demonstrated that the developed polymer/ionic liquid thermoplastic solid electrolyte is a promising candidate for further development of all-solid-state batteries with relatively lower environmental impact.

4. Conclusions

The polymer/ionic liquid thermoplastic solid electrolyte based on PEO, modified sepiolite (TPGS-S), LiTFSI, and PYR$_{14}$TFSI ionic liquid was successfully prepared using a solvent-free extrusion method beneficial for low cost and environment-friendly solid-state battery mass production. The physical-chemical, electrical, and electrochemical properties of the developed solid electrolyte were comprehensively characterized.

The TPE presented a maximum conductivity 0.5 S·cm^{-1} at 60 °C. The LSV curve showed that the electrolyte is stable up to 4.2 V versus Li/Li$^+$ and possesses an excellent compatibility with lithium metal electrode during more than 1600 hours cycling under comparatively harsh cycling conditions (1 mA·cm^{-2}, 2 mAh·cm^{-2}).

Finally, the developed solid electrolyte demonstrated a promising cycling performance in nonoptimized prototype of truly Li-LiFePO$_4$ solid-state coin cell working under relatively higher charge/discharge C-rates (0.2 C/0.5 C).

Thus, the reported solid electrolyte can be considered as a promising candidate for further development of solid-state batteries with charge cut-off voltage about 4.0 V.

Author Contributions: Conceptualization, P.T., A.K.; Methodology, O.G., E.F., A.K.; Materials design, F.G, N.G., P.T., Investigation, F.G., O.G.; Formal Analysis, O.G., A.K.; Data Curation, O.G., E.F., A.K.; Writing-Original Draft Preparation, P.T., F.G., O.G., E.F., A.K., I.U.; Writing-Review & Editing, P.T., N.G., O.G., E.F., A.K., I.U.

Funding: This study was supported by the Comunidad de Madrid (Project Ref. LIQUORGAS-CM, S2013/MAE-2800). Francisco González is grateful to CONACYT-SENER for the scholarship granted (CVU559770/Registro 297710 and S2013/MAE-2800).

Acknowledgments: Leoncio Garrido is acknowledged for the determination of diffusion coefficients.

Conflicts of Interest: The authors declare no conflict of interest.

References

1. Liu, S.F.; Wang, X.L.; Xie, D.; Xia, X.H.; Gu, C.D.; Wu, J.B.; Tu, J.P. Recent development in lithium metal anodes of liquid-state rechargeable batteries. *J. Alloys Compd.* **2018**, *730*, 135–149. [CrossRef]
2. Cheng, X.B.; Zhang, R.; Zhao, C.Z.; Zhang, Q. Toward safe lithium metal anode in rechargeable batteries: A review. *Chem. Rev.* **2017**, *117*, 10403–10473. [CrossRef] [PubMed]
3. Guo, Y.; Li, H.; Zhai, T. Reviving lithium-metal anodes for next-generation high-energy batteries. *Adv. Mater.* **2017**, *29*, 1700007. [CrossRef] [PubMed]
4. Abada, S.; Marlair, G.; Lecocq, A.; Petit, M.; Sauvant-Moynot, V.; Huet, F. Safety focused modeling of lithium-ion batteries: A review. *J. Power Sources* **2016**, *306*, 178. [CrossRef]
5. Varzi, A.; Raccichini, R.; Passerini, S.; Scrosati, B. Challenges and prospects on the role of solid electrolytes for the revitalization of lithium metal batteries. *J. Mater. Chem. A* **2016**, *4*, 17251–17259. [CrossRef]
6. Perea, A.; Dontigny, M.; Zaghib, K. Safety of solid-state Li metal battery: Solid polymer versus liquid electrolyte. *J. Power Sources* **2017**, *359*, 182–185. [CrossRef]
7. Kim, J.G.; Son, B.; Mukherjee, S.; Schuppert, N.; Bates, A.; Kwon, O.; Choi, M.J.; Chung, H.Y.; Park, S. A review of lithium and non-lithium based solid state batteries. *J. Power Sources* **2015**, *282*, 299–322. [CrossRef]
8. Armand, M. Polymer solid electrolytes—An overview. *Solid State Ionics* **1983**, *9–10*, 745–754. [CrossRef]
9. Monroe, C.; Newman, J. The Impact of Elastic Deformation on Deposition Kinetics at Lithium/Polymer Interfaces. *J. Electrochem. Soc.* **2005**, *152*, A396–A404. [CrossRef]
10. Takeda, Y.; Yamamoto, O.; Imanishi, N. Lithium Dendrite Formation on a Lithium Metal Anode from Liquid, Polymer and Solid Electrolytes. *Electrochemistry* **2016**, *84*, 210–218. [CrossRef]
11. Xue, Z.; He, D.; Xie, X. Poly(ethylene oxide)-based electrolytes for lithium-ion batteries. *J. Mater. Chem. A* **2015**, *3*, 19218–19253. [CrossRef]
12. Ngai, K.S.; Ramesh, S.; Ramesh, K.; Juan, J.C. A review of polymer electrolytes: fundamental, approaches and applications. *Ionics* **2016**, *22*, 1259–1279. [CrossRef]
13. Soo, P.P.; Huang, B.; Jang, Y.I.; Chiang, Y.M.; Sadoway, D.R.; Mayesz, A.M. Rubbery Block Copolymer Electrolytes for Solid-State Rechargeable Lithium Batteries. *J. Electrochem. Soc.* **1999**, *146*, 32–37. [CrossRef]
14. Marcinek, M.; Syzdek, J.; Marczewski, M.; Piszcz, M.; Niedzicki, L.; Kalita, M.; Plewa-Marczewska, A.; Bitner, A.; Wieczorek, P.; Trzeciak, T.; et al. Electrolytes for Li-ion transport—Review. *Solid State Ionics* **2015**, *276*, 107–126. [CrossRef]
15. Manuel Stephan, A.; Nahm, K.S. Review on composite polymer electrolytes for lithium batteries. *Polymer* **2006**, *47*, 5952–5964. [CrossRef]
16. Croce, F.; Appetecchi, G.B.; Persi, L.; Scrosati, B. Nanocomposite polymer electrolytes for lithium batteries. *Nature* **1998**, *394*, 456–458. [CrossRef]
17. Weston, J.E.; Steele, B.C.H. Effects of inert fillers on the mechanical and electrochemical properties of lithium salt-poly(ethylene oxide) polymer electrolytes. *Solid State Ionics* **1982**, *7*, 75–79. [CrossRef]
18. Bishop, A.G.; MacFarlane, D.R.; McNaughton, D.; Forsyth, M. FT-IR Investigation of Ion Association in Plasticized Solid Polymer Electrolytes. *J. Phys. Chem. A* **1996**, *100*, 2237–2243. [CrossRef]
19. Piana, M.; Wandt, J.; Meini, S.; Buchberger, I.; Tsiouvaras, N.; Gasteigera, H.A. Stability of a Pyrrolidinium-Based Ionic Liquid in Li-O_2 Cells. *J. Electrochem. Soc.* **2014**, *161*, A1992–A2001. [CrossRef]

20. Osada, I.; de Vries, H.; Scrosati, B.; Passerini, S. Ionic-liquid-based polymer electrolytes for battery applications. *Angew. Chem. Int. Ed.* **2016**, *55*, 500–513. [CrossRef] [PubMed]
21. Ye, Y.S.; Rick, J.; Hwag, B.J. Ionic liquid polymer electrolytes. *J. Mater. Chem. A* **2013**, *1*, 2719–2743. [CrossRef]
22. Shaplov, A.S.; Marcilla, R.; Mecerreyes, D. Recent advances in innovative polymer electrolytes based on poly(ionic liquid)s. *Electrochim. Acta* **2015**, *175*, 18–34. [CrossRef]
23. Yim, T.; Kwon, M.S.; Mun, J.; Kyu, T.L. Room temperature ionic liquid-based electrolytes as an alternative to carbonate-based electrolytes. *Israel J. Chem.* **2015**, *55*, 586–598. [CrossRef]
24. Hirota, N.; Okuno, K.; Majima, M.; Hosoe, A.; Uchida, S.; Ishikawa, M. High-performance lithium-ion capacitor composed of electrodes with porous three-dimensional current collector and bis(fluorosulfonyl) imide-based ionic liquid electrolyte. *Electrochim. Acta* **2018**, *276*, 125–133. [CrossRef]
25. Simonetti, E.; Carawska, M.; Maresca, G.; De Francesco, M.; Appetecchi, G.B. Highly conductive ionic liquid-based polymer electrolytes. *J. Electrochem. Soc.* **2017**, *164*, A6213–A6219. [CrossRef]
26. González, F.; Gregorio, V.; Rubio, A.; Garrido, L.; García, N.; Tiemblo, P. Ionic liquid-based thermoplastic solid electrolytes processed by solvent-free procedures. *Polymers* **2018**, *10*, 124. [CrossRef]
27. Schnell, J.; Günther, T.; Knoche, T.; Vieider, C.; Köhler, L.; Just, A.; Keller, M.; Passerini, S.; Reinhart, G. All-solid-state lithium-ion and lithium metal batteries—paving the way to large-scale production. *J. Power Sources* **2018**, *382*, 160–175. [CrossRef]
28. Mejía, A.; Benito, E.; Guzmán, J.; Garrido, L.; García, N.; Hoyos, M.; Tiemblo, P. Polymer/Ionic Liquid Thermoplastic Electrolytes for Energy Storage Processed by Solvent Free Procedures. *ACS Sustain. Chem. Eng.* **2016**, *4*, 2114–2121. [CrossRef]
29. Pozyczka, K.; Marzantowicz, M.; Dygas, J.R.; Krok, F. Ionic conductivity and lithium transference number of poly (ethylene oxide): LiTFSI system. *Electrochim. Acta* **2017**, *227*, 127–135. [CrossRef]
30. Mejía, A.; García, N.; Guzmán, J.; Tiemblo, P. Surface modification of sepiolite nanofibers with PEG based compounds to prepare polymer electrolytes. *Appl. Clay Sci.* **2014**, *95*, 265–274. [CrossRef]
31. Homminga, D.; Goderis, B.; Dolbnya, I.; Reynaers, H.; Groeninckx, G. Crystallization behavior of polymer/montmorillonite nanocomposites. Part I. Intercalated poly (ethylene oxide)/montmorillonite nanocomposites. *Polymer* **2005**, *46*, 11359–11365. [CrossRef]
32. Borodin, O. Polarizable force field development and molecular dynamics simulations of ionic liquids. *J. Phys. Chem. B* **2009**, *113*, 11463–11478. [CrossRef] [PubMed]
33. Porcarelli, L.; Gerbaldi, C.; Bella, F.; Nair, J.R. Super Soft All-Ethylene Oxide Polymer Electrolyte for Safe All-Solid Lithium Batteries. *Sci. Rep.* **2016**, *6*, 19892. [CrossRef] [PubMed]
34. Evans, J.; Vincent, C.A.; Bruce, P.G. Electrochemical measurement of transference number in polymer electrolytes. *Polymer* **1987**, *28*, 2324–2328. [CrossRef]
35. Martinelli, A.; Matic, A.; Jacobsson, P.; Borjesson, L.; Fernicola, A.; Scrosati, B. Phase Behavior and Ionic Conductivity in Lithium Bis(trifluoromethanesulfonyl)imide-Doped Ionic Liquids of the Pyrrolidinium Cation and Bis(trifluoromethanesulfonyl)imide Anion. *J. Phys. Chem. B* **2009**, *113*, 11245–11251. [CrossRef] [PubMed]

© 2018 by the authors. Licensee MDPI, Basel, Switzerland. This article is an open access article distributed under the terms and conditions of the Creative Commons Attribution (CC BY) license (http://creativecommons.org/licenses/by/4.0/).

Article

Ionic Liquid-Based Electrolyte Membranes for Medium-High Temperature Lithium Polymer Batteries

Guk-Tae Kim [1,2], Stefano Passerini [1,2], Maria Carewska [3] and Giovanni Battista Appetecchi [4,*]

1. Helmholtz Institute Ulm-Karlsruhe Institute of Technology, Helmholtzstrasse 11, 89081 Ulm, Germany; guk-tae.kim@kit.edu (G.-T.K.); stefano.passerini@kit.edu (S.P.)
2. Karlsruher Institute of Technology (KIT), P.O. Box 3640, 76021 Eggenstein-Leopoldshafen, Germany
3. ENEA, Agency for New Technologies, Energy and Sustainable Economic Development, DTE-PCU-SPCT, Via Anguillarese 301, 00123 Rome, Italy; maria.carewska@enea.it
4. ENEA, Agency for New Technologies, Energy and Sustainable Economic Development, SSPT-PROMAS-MATPRO, Via Anguillarese 301, 00123 Rome, Italy
* Correspondence: gianni.appetecchi@enea.it; Tel.: +39-06-3048-3924

Received: 11 June 2018; Accepted: 9 July 2018; Published: 10 July 2018

Abstract: Li^+-conducting polyethylene oxide-based membranes incorporating *N*-butyl-*N*-methylpyrrolidinium bis(trifluoromethanesulfonyl)imide are used as electrolyte separators for all-solid-state lithium polymer batteries operating at medium-high temperatures. The incorporation of the ionic liquid remarkably improves the thermal, ion-transport and interfacial properties of the polymer electrolyte, which, in combination with the wide electrochemical stability even at medium-high temperatures, allows high current rates without any appreciable lithium anode degradation. Battery tests carried out at 80 °C have shown excellent cycling performance and capacity retention, even at high rates, which are never tackled by ionic liquid-free polymer electrolytes. No dendrite growth onto the lithium metal anode was observed.

Keywords: ionic liquids; *N*-butyl-*N*-methylpyrrolidinium bis(trifluoromethanesulfonyl)imide; poly(ethyleneoxide); polymer electrolytes; lithium polymer batteries

---

1. Introduction

Large-scale applications such as automotive, stationary, deep-sea drilling devices need batteries to be capable of operating safely at medium-high temperatures with very good performance and cycle life; i.e., without appreciable degradation phenomena. In addition, even devices generally operating around room temperature could be accidentally subjected to prolonged overheating, thus requiring high thermal stability. In this scenario, electrolytes play a key role.

Rechargeable lithium batteries are an excellent choice as advanced electrochemical energy storage systems due to their high energy density and cycle life [1,2]. Recently published manuscripts report that ion conducting polymer membranes, realized through common materials and up-scalable processes, can act as electrolyte separators for rechargeable lithium battery systems. J.R. Nair et al. [3] have prepared methacrylic-based PEs, reinforced with both cellulose hand-sheets and nanosize cellulose fibers, by UV-induced free radical photo-polymerization. Similarly, rigid–flexible composite electrolyte membranes, based on poly(ethyl α-cyanoacrylate) and cellulose backbone, have been prepared through an in-situ polymerization process by P. Hu et al. [4]. These cross-linking techniques, also successfully proposed for Na^+ conducting PEs [5], have shown short processing times, easy up-scalability and eco-compatibility, and have enabled gel polymer electrolytes (GPEs) with wide electrochemical stability windows and high room temperature ionic conductivity in combination with good mechanical properties to be obtained. Poly(vinylidene difluoride)-based GPEs were obtained via the phase-inversion method [6]. The use of nano-clay filler and pore-forming agent, i.e., poly(vinylpyrrolidone), was seen to significantly improve the electrolyte uptake and the ion transport properties. H. Li et al. [7] have combined the advantages of GPEs with those of ceramic conductors to prepare sandwiched structure composite electrolytes with enhanced electrochemical performance. Reviews of GPE systems, addressed to Li/S [8] and Li-ion [9] battery systems, were recently published.

However, commercial lithium-ion batteries, even employing GPEs, do not behave well at medium-high temperatures as the organic electrolyte quickly degrades above 50 °C, thus irreversibly ageing the electrochemical device [10–12]. In this scenario, the development of solvent-free polymer electrolytes is undoubtedly appealing from safety and engineering points of view and opens new perspectives to applications in electrochemical devices [8,9,13–17]. In addition, polymer electrolytes (PEs) can be easily and cheaply manufactured into low thicknesses (<100 μm) and shapes not allowed for supported liquid electrolytes, offering a new concept of solvent-free, all-solid-state, thin-layer, flexible (both mechanically and in design), robust, lithium polymer batteries (LPBs). Finally, PEs play a second role in composite electrodes as binders and ionic conductors [18].

Nevertheless, the realization of all-solid-state lithium battery systems has been prevented so far by the low ionic conductivity of PEs, especially at ambient temperature. For instance, poly(ethyleneoxide)-lithium salt (PEO-LiX) complexes, considered to be very good candidates as electrolyte separators for LPBs [13–23], approach conduction values of interest for practical applications (>10^{-4} S·cm^{-1}) only above 70 °C, i.e., when the polymer is in the amorphous state [13,14,17,20–22]. However, even at medium-high temperatures (≥90 °C) LPBs exhibit high performance only at low current rates (≤0.1C) [18,22,23], thus preventing applications requiring high power density.

An appealing way to overcome the conductivity drawback is represented by the incorporation of ionic liquids (ILs) into the polymer electrolytes [24]. ILs, i.e., salts which are molten at room temperature consisting of organic cations and inorganic/organic anions [25–27], display several peculiarities such as their extremely low flammability, negligible vapor pressure, high chemical–electrochemical–thermal stability, fast ion transport properties, good power solvency and high specific heat. In the last years, it was successfully demonstrated [24,28–34] how the addition of ILs to PEO-based electrolytes enhances the ionic conductivity above 10^{-4} S·cm^{-1} at 20 °C—i.e., more than two orders of magnitude higher than that of ionic liquid-free PEs—allowing LPBs to obtain a significant cycling performance at near room temperature (30–40 °C) [24,29–34].

In the present work, we show how the incorporation of ionic liquids improves the performance of PEO-based electrolytes even at medium-high temperatures, especially at high current rates, without any evident material degradation and battery cycle life depletion, making the IL-containing PEO membrane an appealing electrolyte separator for LIBs operating at medium-high temperatures. N-butyl-N-methylpyrrolidinium bis(trifluoromethanesulfonyl)imide (PYR$_{14}$TFSI) was selected as the ionic liquid [24].

2. Materials and Methods

2.1. Synthesis of the Ionic Liquid

The PYR$_{14}$TFSI ionic liquid was synthesized through an eco-friend route, reported in detail elsewhere [35,36].

2.2. Preparation of the Polymer Electrolyte and the Composite Cathode

The ionic liquid-based polymer electrolyte and composite cathode were prepared through a solvent-free process [33] carried out in a very low relative humidity dry-room (R.H. < 0.1% at 20 °C). The material components, i.e., PEO (Dow Chemical, Midland, MI, USA, WSR 301, M_W = 4,000,000 a.u.), lithium bis(trifluoromethanesulfonyl)imide (LiTFSI, 3M, battery grade) and PYR$_{14}$TFSI, were vacuum dried at 50 °C for 48 h (PEO) and at 120 °C for 24 h (lithium salt and ionic liquid). PEO and LiTFSI

(EO:Li mole ratio = 1:0.1) were intimately mixed in a mortar, and then PYR$_{14}$TFSI was added to achieve a (PYR$_{14}$)$^+$/Li$^+$ mole ratio equal to 1:1. In previous papers [24,33], we have shown that this ratio represents a good compromise between ion transport properties and interfacial stability. The P(EO)$_1$(LiTFSI)$_{0.1}$(PYR$_{14}$TFSI)$_{0.1}$ past-like electrolyte blend was annealed under vacuum at 100 °C overnight in order to allow the full diffusion of the lithium salt and ionic liquid through the PEO host, therefore obtaining a homogeneous mixture. Finally, the so-obtained rubber-like material was hot-pressed at 100 °C for 2 min to form 70–80 µm thick films. Ionic liquid-free, P(EO)$_1$(LiTFSI)$_{0.1}$ binary polymer electrolytes were prepared for comparison purposes.

The cathode tape was prepared by intimately blending LiFePO$_4$ active material (Sud Chemie, Munich, Germany) and KJB carbon (electronic conductor, Akzo Nobel, Amsterdam, The Netherlands). LiFePO$_4$ and KJB were previously vacuum dried at 120 °C for at least 24 h. Separately, PEO, LiTFSI and PYR$_{14}$TFSI were roughly mixed (to obtain a paste-like mixture) and then added to the LiFePO$_4$-KJB blend. The resulting cathodic mixture was firstly annealed at 100 °C overnight and then hot-pressed to form preliminary films (200–300 µm thick) which were cold-rolled to obtain the final cathode tape (<50 µm) and to remove any porosity within the composite cathode [37]. Finally, 12 mm diameter cathode discs (active area equal to 1.13 cm^2) were punched for the battery tests. The active material mass loading ranged from 4 to 5 mg·cm^{-2}, corresponding (accounting for a theoretical capacity of LiFePO$_4$ equal to 170 mA·h·g^{-1}) to a capacity from 0.7 to 0.8 mA·h·cm^{-2}.

2.3. Thermal Analysis

DSC measurements were run using a differential scanning calorimeter (TA Instruments, model Q100, New Castle, DE, USA). The samples, upon housing (within the dry room) in sealed Al pans, were cooled (10 °C·min^{-1}) from room temperature down to −140 °C and then heated (10 °C·min^{-1}) up to 150 °C.

The thermal stability was verified in a nitrogen atmosphere through TG analysis carried out by a SDT 2960 equipment, simultaneous TG-DTA (TA Instruments, New Castle, DE, USA) with Thermal Solution Software (version 1.4, Thermal Solutions Inc, Ann Arbor, MI, USA). During the experiments, the atmosphere above the samples was fixed by flowing high purity nitrogen atmosphere at a flow rate of 100 mL·min^{-1}. The experiments were performed on 5–10 mg samples (handled in the dry room), which were housed in platinum crucibles. The thermal stability was initially investigated by running a heating scan from room temperature up to 500 °C at a scan rate of 10 °C·min^{-1}.

2.4. Cell Assembly

The electrochemical measurements on the polymer electrolyte samples were carried out on two-electrode cells fabricated in the dry room. Two different cell types (active area equal from 2 to 3 cm^2) were assembled by sandwiching a polymer electrolyte separator between (i) two Li foil electrodes (50 µm thick, supported onto Cu grids as the current collectors) for determining, respectively, the resistance at the interface with the lithium anode and the limiting diffusion current density; (ii) a nickel foil (working electrode, 100 µm thick, used also as the current collector) and a lithium foil (counter electrode, 50 µm thick, supported onto a Cu grid as the current collector) for the linear sweep voltammetry tests. In the latter kind of cell, a tiny lithium strip (50 µm thick, supported onto a Ni grid as the current collector) was used as the reference electrode.

The electronic conductivity of the ionic liquid-containing LiFePO$_4$ composite cathode was investigated as a function of the carbon content by carrying out impedance measurements on symmetrical Al/cathode/Al cells. The composite cathode tape was interlayered between two Al foils (20 µm thick), which were also used as the current collectors.

The solid-state Li/LiFePO$_4$ batteries (cathode limited) were fabricated (inside the dry room) by laminating a lithium foil (50 µm thick), a P(EO)$_{10}$(LiTFSI)$_{0.1}$(PYR$_{14}$TFSI)$_{0.1}$ polymer electrolyte separator and a LiFePO$_4$-based composite cathode tape (plated onto a 20 µm thick Al foil). Aluminum and copper grids were used as the cathodic and anodic current collector, respectively. The electrochemically active area of the Li/LiFePO$_4$ cells was 1.13 cm^2.

All assembled cells were housed in soft envelopes, evacuated for at least 1 h (10^{-2} mbar) and then vacuum-sealed. Finally, the cells were laminated twice by hot-rolling at 100 °C to improve the electrolyte/electrode interfacial contact.

2.5. Electrochemical Tests

Impedance measurements were performed on symmetrical Li/polymer electrolyte/Li (frequency range: 65 kHz–10 mHz; temperature range: 20–80 °C) and Al/composite cathode/Al (10 kHz–1 Hz, 20 °C) cells by a Frequency Response Analyzer, F.R.A. (Schlumberger Solartron, mod. 1260, Leicester, UK). The analysis of the AC responses was carried out by an equivalent circuit model taking into account all possible contributes to the impedance of the cell under test [38]. The validity of the selected circuit was confirmed by fitting the AC responses using a non-linear least-square (NLLSQ) software developed by Boukamp [39,40] (only fits characterized by a χ^2 factor lower than 10^{-4} were considerable acceptable [39,40]).

The electrochemical stability window (ESW) of the P(EO)$_1$(LiTFSI)$_{0.1}$(PYR$_{14}$TFSI)$_{0.1}$ polymer electrolyte was evaluated by linear sweep voltammetries (LSVs) run at 0.5 mV·s^{-1} in the 20–80 °C temperature range. The measurements were performed by scanning the cell potential from the open circuit value (OCV) towards more negative or positive potentials to determine the cathodic and anodic electrochemical stability limits, respectively. The LSVs were performed at least twice on each electrolyte to confirm the results obtained, using fresh samples and clean electrodes for each test. The measurements were performed at 20 °C using an Electrochemical Interface (Schlumberger Solartron, mod. 1287, Leicester, UK).

The limiting diffusion current density of the P(EO)$_1$(LiTFSI)$_{0.1}$ and P(EO)$_1$(LiTFSI)$_{0.1}$(PYR$_{14}$TFSI)$_{0.1}$ polymer electrolytes was determined by potentiodynamic measurements on symmetrical Li/electrolyte/Li cells, i.e., the cell voltage was linearly increased from the OCV value (a few mV) at a scan rate of 0.01 mV·s^{-1} until the current response achieves a steady state. The measurements were performed at temperatures ranging from 40 to 80 °C by a potentiostat/galvanostat (MACCOR, mod. 4000, Tulsa, OK, USA).

The cycling performance of the Li/LiFePO$_4$ polymer cells was evaluated under charge/discharge rates ranging from 0.1C (j = 0.07–0.08 mA·cm^{-2}) to 1C (j = 0.7–0.8 mA·cm^{-2}) at 80 °C. The battery tests were performed using a multiple battery tester (MACCOR, mod. S4000, Tulsa, OK, USA). The voltage cut-offs were fixed at 4.0 V (charge step) and 2.0 V (discharge step), respectively. During the experiments, the cells were held in a climatic chamber (Binder GmbH, mod. MK53, Tuttlingen, Germany) with a temperature control of ±0.1 °C.

3. Results and Discussion

3.1. Ionic Liquid-Based Polymer Electrolytes

The solvent-free procedure allowed homogeneous, freestanding, polymer electrolyte membranes with good mechanical properties to be obtained. In addition, the ionic liquid-containing P(EO)$_1$(LiTFSI)$_{0.1}$(PYR$_{14}$TFSI)$_{0.1}$ sample looks rather sticky, thus resulting (even if not easily handled) in improved contact at the interface with electrodes.

The results of the DSC investigation are illustrated in Figure 1a. The P(EO)$_1$(LiTFSI)$_{0.1}$ electrolyte shows a broad endothermic melting peak centered around 60 °C [21,41] and a weak glass transition (T_g) feature located at −39 °C. The pure PYR$_{14}$TFSI ionic liquid, reported for comparison purposes, exhibits only a melting peak around −7 °C [42]; i.e., the absence of glass transition and exothermal "cold crystallization" features suggest that the IL sample was fully crystallized prior to running the DSC measurements [43]. The incorporation of PYR$_{14}$TFSI into the P(EO)$_1$(LiTFSI)$_{0.1}$ electrolyte results in almost complete disappearance of the melting peak in the DSC trace, which displays only the

T_g feature around −55 °C, clearly indicating that the P(EO)$_1$(LiTFSI)$_{0.1}$(PYR$_{14}$TFSI)$_{0.1}$ electrolyte is amorphous even at room temperature.

The thermal stability is a mandatory requirement for electrolytes to be addressed to battery systems for medium-high temperature applications. Figure 1b compares the TGA trace (in nitrogen atmosphere) of the P(EO)$_1$(LiTFSI)$_{0.1}$ and P(EO)$_1$(LiTFSI)$_{0.1}$(PYR$_{14}$TFSI)$_{0.1}$ electrolyte membranes. The IL-free sample exhibits a weight loss above 180 °C, whereas the addition of the ionic liquid component results in thermal stability increase up to 220 °C. It should be noted that PYR$_{14}$TFSI is seen to be thermally stable up 290 °C. Therefore, we can reasonably hypothesize that the ionic liquid, properly incorporated within the polymer host, is able to protect the PEO chains by thermal degradation. Something similar was previously observed in other PEO electrolytes [41], in which the IL agent, suitably dispersed through the polymeric matrix, was seen to prevent the oxidation of the polymer host above 4 V (vs. Li$^+$/Li°).

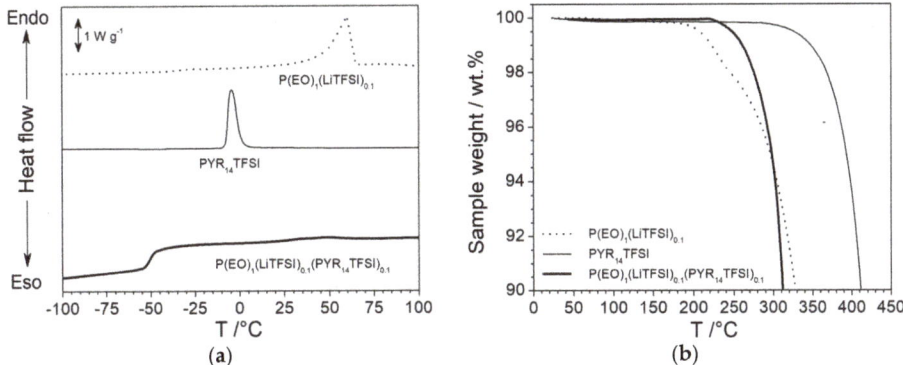

Figure 1. DSC (panel **a**) and TGA (panel **b**) traces of P(EO)$_{10}$(LiTFSI)$_1$ and P(EO)$_{10}$(LiTFSI)$_1$(PYR$_{14}$TFSI)$_1$ polymer electrolyte samples. Scan rate: 10 °C·min^{-1}. The PYR$_{14}$TFSI ionic liquid is reported for comparison purposes.

The effect of the incorporation of the PYR$_{14}$TFSI ionic liquid on the ion transport properties of the polymer electrolyte is summarized in Table 1. A remarkable conductivity increase is observed, especially at ambient temperature and below. For instance, the P(EO)$_1$(LiTFSI)$_{0.1}$(PYR$_{14}$TFSI)$_{0.1}$ sample shows ion conduction values three and two orders of magnitude higher than that of the IL-free sample at −20 °C and 20 °C [31,33], respectively. More than 10^{-4} S·cm^{-1} are exhibited at 20 °C, this is of interest for applications in practical devices and commonly not approached in polymer electrolyte membranes. These results support faster ion conduction through the PEO electrolyte due both to a much larger content of the amorphous phase, in agreement with the DSC data of Figure 1a, and to the enhanced mobility of the Li$^+$ cations resulting from the presence of PYR$_{14}$TFSI; i.e., the addition of ionic liquid results in large anion excess with respect to the lithium cations. Therefore, the strength of the Li$^+$···Anion$^-$ interaction reduces the role of the PEO chains in the coordination of the lithium cations, e.g., as a result from the competition with the PEO···Li$^+$ interactions [24]. At medium-high temperatures, the conductivity of the P(EO)$_1$(LiTFSI)$_{0.1}$(PYR$_{14}$TFSI)$_{0.1}$ electrolyte is seen to approach or exceed 10^{-3} S·cm^{-1}, still displaying a substantial rise with respect to that of the binary IL-free P(EO)$_1$(LiTFSI)$_{0.1}$ [31,33].

Table 1. Ionic conductivity and Li anode/polymer electrolyte interface resistance of the poly(ethyleneoxide) $P(EO))_1(LiTFSI)_{0.1}$ and $P(EO)_1(LiTFSI)_{0.1}(PYR_{14}TFSI)_1$ polymer electrolytes at different temperatures. (*) from ref. [31].

Polymer Electrolyte Sample	Ionic Conductivity/S·cm^{-1}			
	−20 °C	20 °C	50 °C	80 °C
$P(EO)_1(LiTFSI)_{0.1}$ (*)	1.1×10^{-9}	1.3×10^{-6}	2.2×10^{-4}	8.4×10^{-4}
$P(EO)_1(LiTFSI)_{0.1}(PYR_{14}TFSI)_{0.1}$ (*)	9.7×10^{-7}	1.1×10^{-4}	7.9×10^{-4}	1.9×10^{-3}
	Li/PE Interfacial Resistance/cm^2			
$P(EO)_1(LiTFSI)_{0.1}$	n.a.	830 ± 80	82 ± 8	7.0 ± 0.7
$P(EO)_1(LiTFSI)_{0.1}(PYR_{14}TFSI)_{0.1}$	n.a.	750 ± 70	65 ± 6	6.3 ± 0.6

An important requirement for any electrolyte is its capacity to successfully and efficiently allow electrode reactions, at the operating temperature of the device, without appreciable electrochemical degradation (oxidation/reduction) phenomena. Therefore, the electrochemical stability window (ESW) of the $P(EO)_1(LiTFSI)_{0.1}(PYR_{14}TFSI)_{0.1}$ electrolyte system was investigated as a function of the temperature. The results, reported in Figure 2 as linear sweep voltammetry curves, evince only a moderate, even if progressive, reduction of the ESW on passing from 20 to 80 °C. In particular, the anodic stability (related to oxidation processes of the electrolyte) detected at 80 °C differs by just 200 mV with respect to that recorded at 20 °C. Conversely, no practical variation is observed on the cathodic side with the temperature increase, displaying massive electrolyte reduction well below 0 V vs. Li$^+$/Li°, which allows lithium plating also at 80 °C. A very low current flow (<25 µA·cm^{-2}) is observed up to the anodic breakdown voltage, thus supporting the high purity of the $P(EO)_1(LiTFSI)_{0.1}(PYR_{14}TFSI)_{0.1}$ sample. On the cathodic verse, three weak (≤20 µA·cm^{-2}) features, progressively evinced with the temperature increase, are observed around 1.5 V, 0.9 V and 0.5 V vs. Li$^+$/Li°, respectively. Results previously reported in the literature [44] suggest that the peaks located at 1.5 V and 0.5 V vs. Li$^+$/Li° are ascribable to the Li$^+$ cation intercalation process into the native Ni$_x$O film onto the nickel working electrode surface, whereas the feature at 0.9 V is likely due to impurities, i.e., probably water [45]. To summarize, the $P(EO)_1(LiTFSI)_{0.1}(PYR_{14}TFSI)_{0.1}$ electrolyte is allowed to successfully operate at medium-high temperatures.

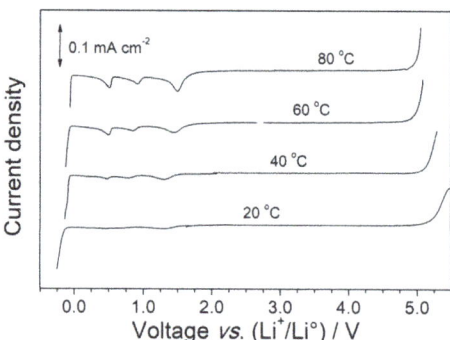

Figure 2. Electrochemical stability window of the $P(EO)_1(LiTFSI)_{0.1}(PYR_{14}TFSI)_{0.1}$ polymer electrolyte sample at different operating temperatures. Nickel is used as the working electrode, lithium as counter and reference electrodes. Scan rate: 0.5 mV·s^{-1}.

The compatibility with the lithium anode is a key parameter for applications as electrolyte separators in Li metal polymer batteries. Figure 3 compares the impedance plots of Li/$P(EO)_1(LiTFSI)_{0.1}$/Li and Li/$P(EO)_1(LiTFSI)_{0.1}(PYR_{14}TFSI)_{0.1}$/Li cells obtained at different temperatures. The AC responses are constituted by a semicircle, taking into account the overall Li/polymer electrolyte interfacial resistance (i.e., charge transfer + passive layer) [38], whereas the high frequency intercept with the real axis is associated with that of the electrolyte bulk [38]. It should be noted that, at 20 °C (panel a), the IL-free electrolyte shows a partial semicircle at high-medium frequencies, due to the relatively low conductivity of the sample $P(EO)_1(LiTFSI)_{0.1}$ [31]. Finally, the inclined straight-line, observed at low frequencies, is attributed to diffusive phenomena through the electrolyte (Warburg contribution) [38]. The impedance plots of Figure 3 clearly confirm how the incorporation of ionic liquid results in a significant decrease of the electrolyte resistance, especially from room to medium temperature, in agreement with the conductivity data reported in Table 1. However, a gain, even if moderate, in interface resistance is also detected. For instance, the $P(EO)_1(LiTFSI)_{0.1}(PYR_{14}TFSI)_{0.1}$ sample shows, at the interface with Li metal, a resistance of 10–11% lower (i.e., from 830 to 750 cm^2 at 20 °C and from 7.0 to 6.3 cm^2 at 80 °C) than that of the IL-free electrolyte (Table 1), in the whole investigated temperature range (20–80 °C). We can hypothesize that the ionic liquid improves the Li$^+$ cation mobility at the electrolyte/lithium interface.

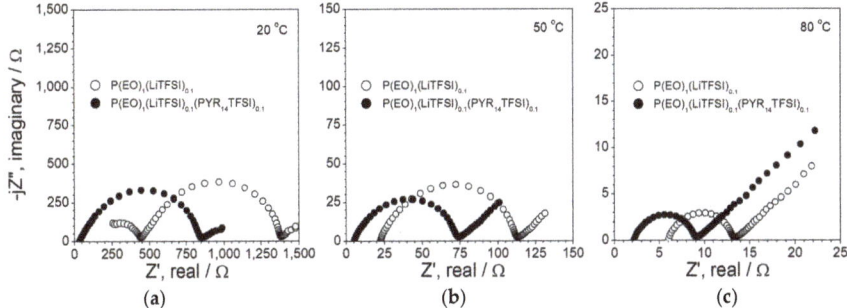

Figure 3. AC response of Li/$P(EO)_1(LiTFSI)_{0.1}$/Li and Li/$P(EO)_1(LiTFSI)_{0.1}(PYR_{14}TFSI)_{0.1}$/Li symmetrical cells at 20 °C (panel **a**), 50 °C (panel **b**) and 80 °C (panel **c**).

Applications such as in automotives, smart grids, etc. require high power and for energy to be readily available; this means that this requires the battery system to be feasibly discharged and charged at high current rates without significantly depleting its performance. For instance, the increase of the current rate promotes the diffusive phenomena within the battery, thus lowering the content of the stored/delivered energy. In electrochemical cells, the redox process kinetics are generally much faster than the active species diffusion through the electrolyte separator. By increasing the current value, the matter transferring process becomes more and more predominant with respect to those at the interfaces with the electrodes. When the current flow through the cell achieves a limiting value, J_L (diffusion limiting current), the electrochemical processes are fully governed by the ion diffusion from the electrolyte bulk to the electrode surface and vice versa. Therefore, J_L is a key parameter for evaluating the feasibility of an electrolyte at high current rates. The limiting current value was determined as reported in Materials and Methods. For instance, linear sweep voltammetry tests were run (at 0.01 mV·s^{-1}) on symmetrical Li/$P(EO)_1(LiTFSI)_{0.1}$/Li and Li/$P(EO)_1(LiTFSI)_{0.1}(PYR_{14}TFSI)_{0.1}$/Li cells at temperatures ranging from 40 to 80 °C. Figure 4 plots the current density values, recorded during the potentiodynamic measurements, as a function of the cell overvoltage. After an initial step increase, in which the electrolyte membrane shows a quasi-ohmic behavior, the current density is seen to progressively level off, likely associated with the establishment of a concentration gradient within the electrolyte membrane [46], around a time-stable value. Such a behavior indicates that the current density through the cell has reached the limiting value (J_L), e.g., the ion conduction processes inside the electrolyte membrane are governed by diffusion phenomena (the concentration gradient extends through the overall electrolyte thickness). In Figure 4, it is shown how the J_L value remarkably increases with the operating cell temperature but is not affected by the presence of $PYR_{14}TFSI$, i.e., from 0.13–017 to 1.2–2.0 mA·cm^{-2} (about one order of magnitude) in passing from 40 to 80 °C for both the IL-free (panel a) and the IL-containing (panel b) electrolyte. Therefore, the ionic liquid does not seem to reduce the diffusive phenomena through the PEO electrolyte. However, the current density of the $P(EO)_1(LiTFSI)_{0.1}$ sample, upon achieving the limiting value, quickly shows an abrupt feature during the potentiodynamic measurements at 60 °C and 80 °C (Figure 4a). This behavior, repeatedly confirmed by several (potentiodynamic) tests carried out on different $Li/P(EO)_1(LiTFSI)_{0.1}/Li$ cells and never observed in the $P(EO)_1(LiTFSI)_{0.1}(PYR_{14}TFSI)_{0.1}$ sample, is ascribable to dendrite growth onto the Li electrode at current rates above 1 mA·cm^{-2}. The results reported in Figure 4a suggest that the IL-free electrolyte is not able to sustain high current rates. Conversely, the ionic liquid plays a key role in improving the compatibility at the interface with the lithium anode, in particular when the cell is subjected to high current rates instead of in an open circuit condition as plotted in Figure 3. It is a plausible hypothesis that $PYR_{14}TFSI$ behaves as a protective agent towards the Li metal electrode, allowing the running of charge/discharge cycling tests at a high current density without appreciable degradation phenomena of the lithium anode. Once more, this confirms the beneficial effect resulting from ionic liquid incorporation on battery performance.

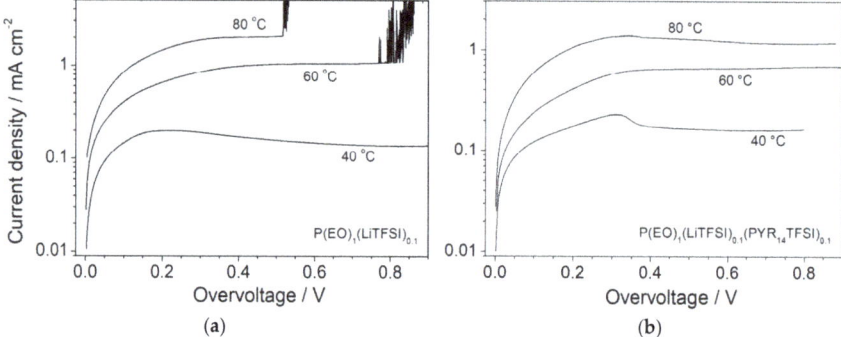

Figure 4. Current density vs. overvoltage curves obtained from potentiodynamic measurements carried out on $Li/P(EO)_1(LiTFSI)_{0.1}/Li$ (panel **a**) and $Li/P(EO)_1(LiTFSI)_{0.1}(PYR_{14}TFSI)_{0.1}/Li$ (panel **b**) cells at different temperatures.

3.2. Composite Electrodes

The LiFePO$_4$ electrode formulation was optimized in terms of carbon content in order to reach a good compromise between electronic conductor content and cathode performance. Therefore, electrode samples containing different carbon contents were prepared and investigated in terms of their electronic conductivity by impedance spectroscopy. The results are reported in Figure 5 as AC responses (panel a) and electronic conductivity vs. carbon content dependence (panel b). The impedance plot of the carbon-free sample (Figure 5a) is constituted by a semicircle (not starting from the axis origin) which does not display any capacitive contribution, indicating charge transfer at the interfaces with the Al° collectors [38]. This behavior—i.e., supporting electron conduction through the composite electrode—suggests the establishment of a three-dimensional network (percolation) formed by LiFePO$_4$ particles and, therefore, electronic continuous pathways through the composite cathode [37]. It should be noted that the as-received active material is provided as superficially carbon-coated; this supports the not-very-low electronic resistance (given by the AC plot intercept with the real axis at low frequencies [38]) of the composite cathode (i.e., pure LiFePO$_4$ material exhibits very low electronic conductivity [47]). The addition of KJB carbon around 3–4 wt. % results in a remarkable reduction of the semicircle diameter and a shifting of the low frequency intercept with the real axis towards smaller impedance values, highlighting a decrease of the electronic resistance of the cathode. At a KJB content equal to 6 wt. %, the semicircle practically reduces to a quasi-single point on the real axis, indicating that the electronic conductivity is largely overcome with respect to the ionic one (the electron and ion conductions through the polymer electrolyte are in parallel) of the polymer electrolyte incorporated within the electrode. In such a condition, the electronic resistance of the composite cathode is given by the distance of the "spot" response intercept with the real axis from the origin of the axes [38].

Figure 5b illustrates the electronic conductivity of the composite LiFePO$_4$ cathode as a function of the carbon content. As evinced in Figure 5a, the electron conduction raises up to 7 wt. % of KJB with a gain of about 1.5 orders of magnitude. The further addition of carbon does not lead to any improvement of the electron transport properties, whereas it depletes the active material content and, therefore, the energy density of the composite cathode. Therefore, the KJB content in the LiFePO$_4$ electrode was fixed to 7 wt. %.

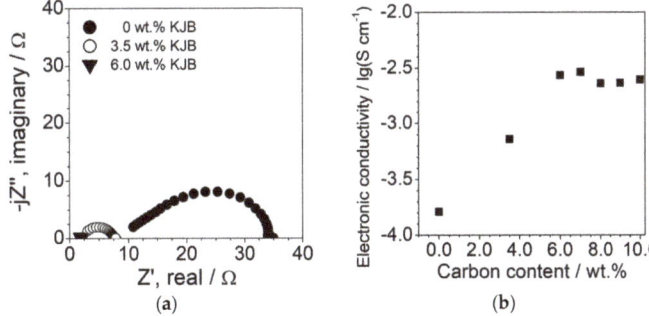

Figure 5. Panel (**a**): impedance plots of Al/LiFePO$_4$ composite cathode/Al symmetrical cells at different carbon contents. Frequency range: 10 kHz–1 Hz. Temperature: 20 °C. Panel (**b**): electronic conductivity of LiFePO$_4$ composite cathode as a function of the carbon content. Temperature: 20 °C.

3.3. Battery Tests at 80 °C

Upon investigation of the electrochemical performance, the P(EO)$_1$(LiTFSI)$_{0.1}$(PYR$_{14}$TFSI)$_{0.1}$ ionic liquid-based, polymer electrolyte was subjected to tests in Li/LiFePO$_4$ cells at 80 °C. Figure 6a compares the voltage vs. capacity profile referring to the 1st charge–discharge cycle run at different current rates. A flat plateau, typical of the Li$^+$ insertion/de-insertion process into the LiFePO$_4$ active material [24,33,34], is observed (in the 3.3–3.6 V range) even at higher rates, with a coulombic efficiency close to 99%. This highlights that IL-incorporating Li/LiFePO$_4$ cells are capable of maintaining the same voltage during almost the entire charge/discharge step. Only a 100 mV increase in ohmic drop is observed on passing from 0.1C through 1C. An initial capacity corresponding to the theoretical value (170 mA·h·g^{-1}) is delivered up to the medium rate (0.33C) with just a moderate decrease at high current rates, i.e., more than 160 mA·h·g^{-1} (>94.1% of the theoretical capacity) are discharged at 1C. Figure 6b,c compares the voltage profiles of the selected charge/discharge cycles at 0.1C and 1C, respectively. It is worth noting that the excellent reproducibility of the battery performance, i.e., the profile feature and the delivered capacity, are practically unchanged after 100 consecutive cycles run (at 100% of deep of discharge, DOD) even at high current rates, which is not often reported for lab-scale, lithium metal polymer cells [24]. These results clearly show the very good reversibility of the Li$^+$ intercalation process even under hard operating conditions in combination with an excellent compatibility at the electrolyte/electrode interface and negligible degradation phenomena occurring within the cell components. Such a performance score, however, can be achieved only through good manufacturing of the electrolyte/electrode components, i.e., high purity levels and careful optimization of the formulation, and of the full cells.

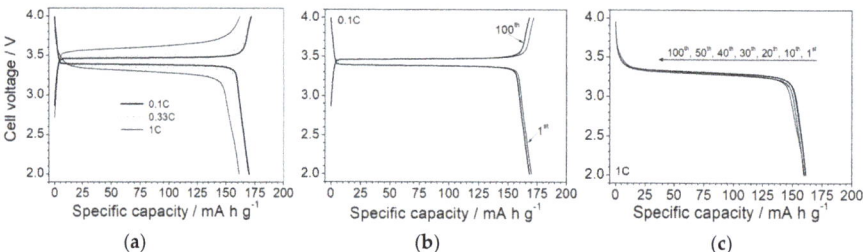

Figure 6. Panel (**a**): voltage vs. charge/discharge capacity profile of the 1st cycle of Li/P(EO)$_1$(LiTFSI)$_{0.1}$(PYR$_{14}$TFSI)$_{0.1}$/LiFePO$_4$ polymeric cells at 80 °C and different current rates. Selected voltage vs. charge/discharge capacity profiles, obtained at 80 °C, of Li/P(EO)$_1$(LiTFSI)$_{0.1}$(PYR$_{14}$TFSI)$_{0.1}$/LiFePO$_4$ cells at 0.1C (panel **b**) and 1C (panel **c**), respectively.

The cycling performance of the Li/P(EO)$_1$(LiTFSI)$_{0.1}$(PYR$_{14}$TFSI)$_{0.1}$/LiFePO$_4$ solid-state cells, tested at 80 °C and different current rates, is depicted in Figure 7a. An excellent capacity retention (as also evinced in Figure 6b,c) with a coulombic efficiency quickly leveling above 99.5% (100% at 0.1C) is recorded even at higher rates, i.e., more than 99.5% and 94% of theoretical capacity are initially delivered at 0.33C and 1C, respectively, with a very modest decay (>98% and 93.6%, respectively) after 100 consecutive cycles. This corresponds to a capacity fading around 0.005% per cycle and, in conjunction with the very good charge/discharge efficiency, once more highlights a highly reversible lithiation process in combination with the high purity level and high compatibility of the P(EO)$_1$(LiTFSI)$_{0.1}$(PYR$_{14}$TFSI)$_{0.1}$ polymer electrolyte towards electrodes, in particular with the lithium metal anode, even at high current rates. Also, it should be noted that very clean lithium metal tapes were used for the cell manufacturing in order to obtain an optimal Li/electrolyte interface. Especially, we would like to point out the absence of dendrite growth on the Li electrode during prolonged cycling tests run also at 1C, i.e., very rarely encountered in lithium metal polymer batteries operating at medium-high temperatures under high rates [24]. These experimental data, in rather good agreement with the results derived from potentiodynamic measurements depicted in Figure 4, once more demonstrate that the incorporation of ionic liquids such as PYR$_{14}$TFSI significantly improves the PEO electrolyte interface with the lithium anode, allowing high current rates to be sustained for prolonged cycling tests without appreciably depleting the cell performance.

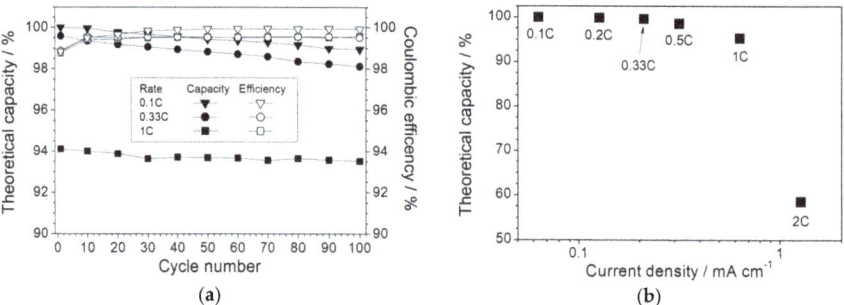

Figure 7. Capacity and coulombic efficiency vs. cycle number evolution at different current rates (panel **a**); and theoretical capacity vs. current density dependence (panel **b**) of Li/P(EO)$_1$(LiTFSI)$_{0.1}$(PYR$_{14}$TFSI)$_{0.1}$/LiFePO$_4$ polymeric cells at 80 °C. The corresponding current rates are also reported.

The capacity vs. current density dependence (80 °C) is plotted in Figure 7b, which evinces a very good rate capability. Above 94% of the theoretical value is still obtained at 1C, supporting an excellent rate capability up to 1C, i.e., corresponding to about 0.7 mA·cm^{-2}, which represents a very interesting current value for an all-solid-state polymer electrolyte. A further increase of the current rate up to 2C, i.e., around 1.4 mA·cm^{-2}, leads to a reduction of the delivered capacity which levels off at 57% of the theoretical value. This behavior, ascribable to diffusive phenomena within the electrolyte separator, is in good agreement with the results obtained by the potentiodynamic measurements (Figure 3b), which indicates that above a current density of about 1.2 mA·cm^{-2} (determined as J_L value), the electrochemical processes through the cell are controlled by the diffusive phenomena occurring within the polymer electrolyte. However, despite the capacity decay due to the operating current density exceeding the limiting value, the Li/P(EO)$_1$(LiTFSI)$_{0.1}$(PYR$_{14}$TFSI)$_{0.1}$/LiFePO$_4$ cells are still able to deliver about 100 mA·h·g^{-1} at a rate as high as 2C (about 1.4 mA·cm^{-2}), i.e., representing a remarkable capacity value for an all-solid-state polymer electrolyte.

The battery performance of the P(EO)$_1$(LiTFSI)$_{0.1}$(PYR$_{14}$TFSI)$_{0.1}$ electrolyte, detected at 80 °C in Li/LiFePO$_4$ cells, is compared with that of other lithium-conducting, ionic liquid-free, PEO membranes, recorded in Li/LiFePO$_4$ and Li/V$_2$O$_5$ systems at temperatures from 90 °C to 100 °C [18,22,23]. The data, reported in Table 2, show how appreciable capacities, i.e., from 70 to 96% of the cell theoretical value, are delivered only at low-medium rates (0.2C–0.33C). However, a capacity decay down to 45–60% of the theoretical value is observed after 100 consecutive charge/discharge cycles, with a fading corresponding to 0.26–0.36% per cycle. Conversely, very modest capacities, i.e., from 8 to 14% of the theoretical value, are obtained when the current rate is increased up to 0.8C–1C. From the data illustrated in Figures 6 and 7 and Table 2, it is evident how, at medium-high temperatures, the PYR$_{14}$TFSI-incorporating lithium polymer batteries behave much better in terms of their delivered capacity and cycling performance than the IL-free ones. For instance, the addition of suitable ionic liquid is able to largely improve the performance of the LPBs not only at ambient or near ambient conditions, as previously reported in the literature [18,20,22,23], but even at medium-high temperatures. Therefore, the PEO-LiTFSI-PYR$_{14}$TFSI Li$^+$-conducting membranes are very promising candidates as electrolyte separator systems for all-solid-state lithium polymer batteries operating around 100 °C.

Table 2. Summary of the battery performance of the P(EO)$_1$(LiTFSI)$_{0.1}$(PYR$_{14}$TFSI)$_1$ polymer electrolyte at 80 °C compared with that of lithium-conducting, ionic liquid-free, PEO membranes at medium-high temperatures. (a) From reference [22]; (b) from reference [23]; (c) from reference [18]; (d) this work.

Polymer Electrolyte Sample	Battery System	T/°C	Current Density/mA·cm^{-2}	Percent of Theoretical Capacity/%
P(EO)$_1$(LiCF$_3$SO$_3$)$_{0.05}$ (a)	Li/Cu$_{0.1}$V$_2$O$_5$	90	0.1 (0.2C)	96 (1st) → 60 (100th)
P(EO)$_1$(LiBETI)$_{0.05}$ (b)	Li/V$_2$O$_5$	90	0.24 (0.33C)	70 (1st) → 45 (100th)
P(EO)$_1$(LiBETI)$_{0.05}$ (b)	Li/V$_2$O$_5$	90	0.72 (1C)	14 (1st)
P(EO)$_1$(LiCF$_3$SO$_3$)$_{0.03}$ + 5 wt. % SiO$_2$ (c)	Li/LiFePO$_4$	100	0.2 (0.2C)	82 (1st) → 47 (100th)
P(EO)$_1$(LiCF$_3$SO$_3$)$_{0.03}$ + 5 wt. % SiO$_2$ (c)	Li/LiFePO$_4$	100	0.8 (0.8C)	8 (1st)
P(EO)$_1$(LiTFSI)$_{0.1}$(PYR$_{14}$TFSI)$_{0.1}$ (d)	Li/LiFePO$_4$	80	0.7 (1C)	94.1 (1st) → 93.6 (100th)

4. Conclusions

PEO-LiTFSI Li$^+$-conducting membranes, containing the PYR$_{14}$TFSI ionic liquid, were prepared and studied to be addressed as electrolyte separators for all-solid-state lithium polymer batteries operating at medium-high temperatures. A solvent-free procedure was designed to prepare the PEO-LiTFSI-PYR$_{14}$TFSI electrolytes. These ternary systems have shown remarkably improved thermal, ion transport and interfacial properties with respect to the ionic liquid-free electrolytes. Wide electrochemical stability was observed even at medium-high operating temperatures. In particular, the ionic liquid-based PEO electrolytes are able to sustain high current rates without any appreciable lithium anode degradation, which is not allowed in binary ionic liquid-free, PEO-LiTFSI systems, thus enabling their use in battery systems operating at 80 °C or above and high current rates. Battery tests carried out at 80 °C in Li/LiFePO$_4$ polymeric systems have shown excellent cycling behavior and capability retention at high current rates, e.g., more than 93.6% of the theoretical capacity (i.e., 99.5% of the initial value) is still delivered after 100 cycles run at 1C with a coulombic efficiency close 100%. This performance largely exceeds that of analogous, ionic liquid-free, polymer lithium batteries at the same operating conditions, nominating the PEO-LiTFSI-PYR$_{14}$TFSI ternary system as an electrolyte separator for medium-high temperature lithium polymer batteries. It is worth highlighting the absence of dendrite growth on the Li anode during prolonged cycling tests even at high current rates, which is very often not observed in lithium metal polymer batteries.

Author Contributions: G.B.A. and S.P. conceived and designed the experiments; G.-T.K. and M.C. performed the experiments; G.-T.K. and G.B.A. analyzed the data; G.B.A. wrote the paper.

Funding: This research received no external funding.

Conflicts of Interest: The authors declare no conflict of interest.

References

1. Notter, D.A.; Gauch, M.; Widmer, R.; Wäger, P.; Stamp, A.; Zah, R.; Althaus, H.-J. Contribution of Li-ion batteries to the environmental impact of electric vehicles. *Environ. Sci. Technol.* **2010**, *44*, 6550–6556. [CrossRef] [PubMed]
2. Yang, H.; Amiruddin, S.; Bang, H.J.; Sun, Y.K.; Prakash, J. A review of Li-ion cell chemistries and their potential use as hybrid electric vehicles. *J. Ind. Eng. Chem.* **2006**, *12*, 12–38.
3. Nair, J.R.; Chiappone, A.; Destro, M.; Jabbour, L.; Meligrana, G.; Gerbaldi, C. UV-induced radical photopolymerization: A smart tool for preparing polymer electrolyte membranes for energy storage devices. *Membranes* **2012**, *2*, 687–704. [CrossRef] [PubMed]
4. Hu, P.; Duan, Y.; Hu, D.; Qin, B.; Zhang, J.; Wang, D.; Liu, Z.; Cui, G.; Chen, L. Rigid−flexible coupling high ionic conductivity polymer electrolyte for an enhanced performance of LiMn$_2$O$_4$/graphite battery at elevated temperature. *ACS Appl. Mater. Int.* **2015**, *7*, 4720–4727. [CrossRef] [PubMed]
5. Colò, F.; Bella, F.; Nair, J.R.; Gerbaldi, C. Light-cured polymer electrolytes for safe, low-cost and sustainable sodium-ion batteries. *J. Power Sources* **2017**, *365*, 293–302. [CrossRef]
6. Dyartanti, E.R.; Purwanto, A.; Widiasa, I.N.; Susanto, H. Ionic conductivity and cycling stability improvement of PVdF/nano-clay using PVP as polymer electrolyte membranes for LiFePO$_4$ batteries. *Membranes* **2018**, *8*, 36. [CrossRef] [PubMed]
7. Lia, H.; Lia, M.; Siyala, S.H.; Zhua, M.; Lana, J.-L.; Suia, G.; Yua, Y.; Zhonga, W.; Yang, X. A sandwich structure polymer/polymer-ceramics/polymer gel electrolytes for the safe, stable cycling of lithium metal batteries. *J. Membr. Sci.* **2018**, *555*, 169–176. [CrossRef]
8. Zhao, Y.; Zhang, Y.; Gosselink, D.; Long Doan, T.N.; Sadhu, M.; Cheang, H.J.; Chen, P. Polymer electrolytes for lithium/sulfur batteries. *Membranes* **2012**, *2*, 553–564. [CrossRef] [PubMed]
9. Yang, M.; Hou, J. Membranes in lithium ion batteries. *Membranes* **2012**, *2*, 367–383. [CrossRef] [PubMed]
10. Spotnitz, R.; Franklin, J. Abuse behavior of high-power, lithium-ion cells. *J. Power Sources* **2003**, *113*, 81–100. [CrossRef]
11. Abraham, D.P.; Roth, E.P.; Kostecky, R.; McCarthy, K.; MacLaren, S.; Doughty, D.H. Diagnostic examination of thermally abused high-power lithium-ion cells. *J. Power Sources* **2006**, *161*, 648–657. [CrossRef]
12. Bandhauer, T.M.; Garimella, S.; Fuller, T.F. A critical review of thermal issues in lithium-ion batteries. *J. Electrochem. Soc.* **2011**, *158*, R1–R25. [CrossRef]
13. Armand, M.; Chabagno, J.M.; Duclot, M. Poly-ethers as solid electrolytes. In *Fast Ion Transport in Solids. Electrodes and Electrolytes*; Vashitshta, P., Mundy, J.N., Shenoy, G.K., Eds.; North Holland Publishers: Amsterdam, The Netherlands, 1979.
14. Gray, F.M. *Polymer Electrolytes*; Royal Society of Chemistry Monographs: Cambridge, UK, 1997.
15. Lightfoot, P.; Metha, M.A.; Bruce, P.G. Crystal structure of the polymer electrolyte Poly(ethylene oxide)$_3$: LiCF$_3$SO$_3$. *Science* **1993**, *262*, 883–885. [CrossRef] [PubMed]
16. Vincent, C.A.; Scrosati, B. *Modern Batteries. An Introduction to Electrochemical Power Sources*, 2nd ed.; Arnold: London, UK, 1993.

17. Gray, F.M.; Armand, M.; Osaka, T. *Energy Storage System for Electronics*; Osaka, T., Datta, M., Eds.; Gordon and Breach Science Publications: Amsterdam, The Netherlands, 2000.
18. Appetecchi, G.B.; Croce, F.; Hassoun, J.; Scrosati, B.; Salomon, M.; Cassel, F. Hot-pressed, solvent-free, nanocomposite, PEO-based electrolyte membranes. II. All-solid, Li/LiFePO$_4$ polymer batteries. *J. Power Sources* **2003**, *124*, 246–253. [CrossRef]
19. Appetecchi, G.B.; Scaccia, S.; Passerini, S. Investigation on the stability of the lithium-polymer electrolyte interface. *J. Electrochem. Soc.* **2000**, *147*, 4448–4452. [CrossRef]
20. Appetecchi, G.B.; Alessandrini, F.; Duan, R.G.; Arzu, A.; Passerini, S. Electrochemical testing of industrially produced PEO-based polymer electrolytes. *J. Power Sources* **2001**, *101*, 42–46. [CrossRef]
21. Appetecchi, G.B.; Henderson, W.; Villano, P.; Berrettoni, M.; Passerini, S. PEO-LiN(SO$_2$CF$_2$CF$_3$)$_2$ polymer electrolytes. I. XRD, DSC and ionic conductivity characterization. *J. Electrochem. Soc.* **2001**, *148*, 1171–1178. [CrossRef]
22. Appetecchi, G.B.; Alessandrini, F.; Carewska, M.; Caruso, T.; Prosini, P.P.; Scaccia, S.; Passerini, S. Investigation on the lithium polymer electrolyte batteries. *J. Power Sources* **2001**, *97*, 790–794. [CrossRef]
23. Villano, P.; Carewska, M.; Appetecchi, G.B.; Passerini, S. PEO-LiN(SO$_2$CF$_2$CF$_3$)$_2$ polymer electrolytes. III. Tests in batteries. *J. Electrochem. Soc.* **2002**, *149*, A1282–A1285. [CrossRef]
24. Passerini, S.; Montanino, M.; Appetecchi, G.B. Lithium polymer batteries based on ionic liquids. In *Polymers for Energy Storage and Conversion*; Mittal, V., Ed.; John Wiley and Scrivener Publishing: Beverly, MA, USA, 2013.
25. Chiappe, C.; Pieraccini, D. Ionic liquids: Solvent properties and organic reactivity. *J. Phys. Org. Chem.* **2005**, *18*, 275–297. [CrossRef]
26. Rogers, J.R.D.; Seddon, K.R. *Ionic Liquids: Industrial Application to Green Chemistry*; ACS Symposium Series 818; American Chemical Society: Washington, DC, USA, 2002.
27. Ohno, H. *Electrochemical Aspects of Ionic Liquids*; John Wiley & Sons Inc.: Hoboken, NJ, USA, 2005.
28. Shin, J.-H.; Henderson, W.A.; Passerini, S. Ionic liquids to the rescue? Overcoming the ionic conductivity limitations of polymer electrolytes. *Electrochem. Commun.* **2003**, *5*, 1016–1020. [CrossRef]
29. Shin, J.-H.; Henderson, W.A.; Appetecchi, G.B.; Alessandrini, F.; Passerini, S. Recent developments in the ENEA lithium metal battery project. *Electrochim. Acta* **2005**, *50*, 3859–3865. [CrossRef]
30. Shin, J.-H.; Henderson, W.A.; Tizzani, C.; Passerini, S.; Jeong, S.-S.; Kim, K.-W. Characterization of solvent-free polymer electrolytes consisting of ternary PEO-LiTFSI-PYR$_{14}$TFSI. *J. Electrochem. Soc.* **2006**, *153*, A1649–A1654. [CrossRef]
31. Kim, G.-T.; Appetecchi, G.B.; Alessandrini, F.; Passerini, S. Solvent-free, PYR1ATFSI ionic liquids-based ternary polymer electrolyte systems. I. Electrochemical characterization. *J. Power Sources* **2007**, *171*, 861–869. [CrossRef]
32. Kim, G.-T.; Appetecchi, G.B.; Carewska, M.; Joost, M.; Balducci, A.; Winter, M.; Passerini, S. UV cross-linked, lithium-conducting ternary polymer electrolytes containing ionic-liquids. *J. Power Sources* **2010**, *195*, 6130–6137. [CrossRef]
33. Appetecchi, G.B.; Kim, G.-T.; Montanino, M.; Alessandrini, F.; Passerini, S. Room temperature lithium polymer batteries based on ionic liquids. *J. Power Sources* **2011**, *196*, 6703–6709. [CrossRef]
34. Kim, G.-T.; Jeong, S.-S.; Xue, M.-Z.; Balducci, A.; Winter, M.; Passerini, S.; Alessandrini, F.; Appetecchi, G.B. Development of ionic liquid-based lithium battery prototypes. *J. Power Sources* **2012**, *199*, 239–246. [CrossRef]
35. Montanino, M.; Alessandrini, F.; Passerini, S.; Appetecchi, G.B. Water-based synthesis of hydrophobic ionic liquids for high-energy electrochemical devices. *Electrochim. Acta* **2013**, *96*, 124–133. [CrossRef]
36. De Francesco, M.; Simonetti, E.; Giorgi, G.; Appetecchi, G.B. About purification route of hydrophobic ionic liquids. *Challenges* **2017**, *8*, 11. [CrossRef]
37. Appetecchi, G.B.; Carewska, M.; Alessandrini, F.; Prosini, P.P.; Passerini, S. Characterization of PEO-based composite cathodes. I. Morphological, thermal, mechanical and electrical properties. *J. Electrochem. Soc.* **2000**, *147*, 451–459. [CrossRef]
38. MacDonald, J.R. *Impedance Spectroscopy*; John Wiley & Sons: New York, NY, USA, 1987.
39. Boukamp, B.A. A package for impedance/admittance data analysis. *Solid State Ion.* **1986**, *18*, 136–140. [CrossRef]
40. Boukamp, B.A. A nonlinear least squares fit procedure for analysis of immittance data of electrochemical systems. *Solid State Ion.* **1986**, *20*, 31–44. [CrossRef]

41. Simonetti, E.; Carewska, M.; Di Carli, M.; Moreno, M.; De Francesco, M.; Appetecchi, G.B. Towards improvement of the electrochemical properties of ionic liquid-containing polyethylene oxide-based electrolytes. *Electrochim. Acta* **2017**, *235*, 323–331. [CrossRef]
42. Appetecchi, G.B.; Montanino, M.; Carewska, M.; Moreno, M.; Alessandrini, F.; Passerini, S. Chemical-physical properties of bis(perfluoroalkylsulfonyl)imide anion-based ionic liquids. *Electrochim. Acta* **2011**, *56*, 1300–1307. [CrossRef]
43. Henderson, W.A.; Passerini, S. Phase behavior of ionic liquid—LiX mixtures: pyrrolidinium cations and TFSI⁻ anions. *Chem. Mater.* **2004**, *16*, 2881–2885. [CrossRef]
44. Passerini, S.; Scrosati, B. Characterization of nonstoichiometric nickel oxide thin-film electrodes. *J. Electrochem. Soc.* **1994**, *141*, 889–895. [CrossRef]
45. Randstrom, S.; Montanino, M.; Appetecchi, G.B.; Lagergren, C.; Moreno, A.; Passerini, S. Effect of water and oxygen traces on the cathodic stability of *N*-alkyl-*N*-methylpyrrolidinium bis(trifluoromethanesulfonyl)imide. *Electrochim. Acta* **2008**, *53*, 6397–6401. [CrossRef]
46. Bard, A.J.; Faulkner, L.R. *Electrochemical Methods*; Wiley: New York, NY, USA, 1980.
47. Wang, C.; Hong, J. Ionic/electronic conducting characteristics of LiFePO$_4$ cathode materials. The determining factors for high rate performance. *J. Electrochem. Soc.* **2007**, *10*, A65–A69. [CrossRef]

© 2018 by the authors. Licensee MDPI, Basel, Switzerland. This article is an open access article distributed under the terms and conditions of the Creative Commons Attribution (CC BY) license (http://creativecommons.org/licenses/by/4.0/).

Article

Composite Nafion Membranes with CaTiO$_{3-\delta}$ Additive for Possible Applications in Electrochemical Devices

Lucia Mazzapioda [1], Maria Assunta Navarra [1], Francesco Trequattrini [2], Annalisa Paolone [3], Khalid Elamin [4], Anna Martinelli [4] and Oriele Palumbo [3,*]

1. Department of Chemistry, Sapienza University of Rome, Piazzale Aldo Moro 5, 00185 Rome, Italy; lucia.mazzapioda@uniroma1.it (L.M.); mariassunta.navarra@uniroma1.it (M.A.N.)
2. Department of Physics, Sapienza University of Rome, Piazzale Aldo Moro 5, 00185 Rome, Italy; francesco.trequattrini@roma1.infn.it
3. Consiglio Nazionale delle Ricerche, Istituto dei Sistemi Complessi, U.O.S. La Sapienza, Piazzale A. Moro 5, 00185 Roma, Italy; annalisa.paolone@roma1.infn.it
4. Department of Chemistry and Chemical Engineering, Chalmers University of Technology, 41296 Gothenburg, Sweden; khalid.elamin@chalmers.se (K.E.); anna.martinelli@chalmers.se (A.M.)
* Correspondence: oriele.palumbo@roma1.infn.it; Tel.: +39-06-49914400

Received: 30 September 2019; Accepted: 28 October 2019; Published: 31 October 2019

Abstract: A composite membrane based on a Nafion polymer matrix incorporating a non-stoichiometric calcium titanium oxide (CaTiO$_{3-\delta}$) additive was synthesized and characterized by means of thermal analysis, dynamic mechanical analysis, and broadband dielectric spectroscopy at different filler contents; namely two concentrations of 5 and 10 wt.% of the CaTiO$_{3-\delta}$ additive, with respect to the dry Nafion content, were considered. The membrane with the lower amount of additive displayed the highest water affinity and the highest conductivity, indicating that a too-high dose of additive can be detrimental for these particular properties. The mechanical properties of the composite membranes are similar to those of the plain Nafion membrane and are even slightly improved by the filler addition. These findings indicate that perovskite oxides can be useful as a water-retention and reinforcing additive in low-humidity proton-exchange membranes.

Keywords: Nafion; CaTiO$_{3-\delta}$; inorganic filler; composite electrolyte

1. Introduction

Nafion is the archetypical membrane for use in the proton exchange membrane fuel cell (PEMFC), a clean technology suitable for both transport and stationary applications [1,2]. However, to be more efficient electrical power generators, fuel cells should operate at low relative humidity (RH) conditions and at temperatures higher than 80 °C. Indeed, at these conditions, the kinetics of the electrode reactions, the tolerance to fuel contaminants, such as carbon monoxide, and the ions' transport properties are all improved. Furthermore, an increased operating temperature would mitigate the problems related to thermal and water management [3].

Unfortunately, in PEMFCs operating at temperatures above 80 °C, Nafion experiences a severe decrease in proton conductivity due to water evaporation, which is also reflected in an increase of the electrolyte's ohmic resistance [4].

One strategy to develop PMFC electrolytes suitable for high-temperature operation is to modify the polymer matrix with inorganic additives that are able to improve the water retention capacity of Nafion membranes. These additives, which can be metal oxide nanoparticles, such as SiO$_2$, TiO$_2$, and ZrO$_2$, or functionalized inorganic materials, such as sulfated metal oxide, reinforce the hydration and proton conductivity of the membranes, thanks to their acidic and hygroscopic properties, allowing them to work at the desired conditions described above [5–8]. In particular, the incorporation of hygroscopic particles in the host polymer can increase the accessible surface in Nafion membranes, facilitating water trapping and creating an additional way for the transfer of protons through the membrane, hence improving the overall fuel cell performance at high temperatures.

Titanium-based oxides are very attractive materials, thanks to their good features in terms of cost, stability, and high natural abundance, triggering the R&D toward the improvement of their chemical–physical properties by the study of their chemistry and surface geometry, making them useful to several research fields [9]. In particular, oxides with a perovskite structure are materials that have found the potential for a wide range of applications, such as sensors, optical devices, and solid-oxide-fuel-cells electrodes and electrolytes [10]. This could be explained by the flexibility of the perovskite design (ABO_3), which allows for the accommodation of several doping agents, such as metal transition elements at both A- and/or B-sites in the lattice. This leads to changes of the electrical and optical properties of the oxide, introducing new electronic levels in the energy gap and providing oxygen vacancies in the lattice of the perovskite [11].

It is expected that the main features of perovskites are controlled by the size and nature of both A and B cations in the ABO_3 structure and that the occupation of the B-site by several ions with different acid/base properties affects the stability of the perovskite. Furthermore, the presence of the oxygen vacancies in the structure can play a key role as active sites for the oxygen adsorption and the dissociative absorption of water, for which the protons conductivity is favored. The last aspect could be used to improve the hydrophilicity of the oxide by the protonation of the lattice oxygen ions. This phenomenon seems to be predominant in perovskite-type oxides due to the low formation enthalpies of oxygen ions as a consequence of low bond strengths and strong relaxation effects [12–14].

Consequently, with the aim to combine all the features of the abovementioned materials, a non-stoichiometric perovskite titanium oxide, calcium titanate ($CaTiO_{3-\delta}$), is here proposed as a water-retention and reinforcing additive in low-humidity proton-exchange membranes. The scope of the present study is to investigate the impact of the additive on thermomechanical and proton-conduction properties of membranes by differential scanning calorimetry (DSC), dynamic mechanical analysis (DMA), and broadband dielectric spectroscopy (BDS) studies, in order to evaluate the interplay of the transition/relaxation phenomena in Nafion membranes at high temperatures and under humidified conditions.

2. Materials and Methods

The $CaTiO_{3-\delta}$ additive was prepared by a template-driven procedure recently proposed by our group [14], using Pluronic F127 to control the structure of the particles and to promote the formation of oxygen vacancies. More specifically, titanium isopropoxide and calcium dichloride dehydrated were used as precursors. Pluronic F127 was dissolved in ethanol (molar ratio 1:4), under vigorous stirring, for 20 min, at 60 °C. Subsequently, titanium isopropoxide was added into the above solution, to obtain a titanium-oxide-based sol. At the same time, calcium dichloride was dissolved in deionized water, and, after 15 min, it was added into the mixture. A 3M NaOH solution was dripped to facilitate the complete dissolution of TiO_2 and its conversion to $CaTiO_3$. The solution was transferred in an autoclave and treated at 180 °C, for 24 h, followed by natural cooling to room temperature. After that, the solid product was centrifuged and washed several times with bi-distilled water. Finally, the perovskite was calcinated for 3 h, at 550 °C (heating rate 3 °C/min), to remove the occluded template. During this step, the polymer was oxidized, creating a reductive environmental near the oxide surface and facilitating the formation of oxygen defects. All the reagents for this synthesis were Sigma-Aldrich products (Sigma-Aldrich, St. Louis, MO, USA).

The sample particles obtained have well-defined prismatic, quasi-cubic morphology, as well as the presence of holes/imperfections on the perovskite surface, which can be observed. Based on Brunauer–Emmett–Teller (BET) analysis, the specific surface area was found to be 6.6 ± 0.5 $m^2 \cdot g^{-1}$.

A solvent-casting procedure was used to prepare both doped and undoped Nafion membranes, according to an established procedure [15–17]. Solvents of a commercial Nafion 5 wt.% dispersion (E.W. 1100, Ion Power Inc., München, Germany) were gradually replaced with N,N-dimethylacetamide (>99.5%, Sigma-Aldrich, St. Louis, MO, USA), at 80 °C. For the composite membranes, two filler concentrations of 5 and 10 wt.% of the $CaTiO_{3-\delta}$ additive, with respect to the dry Nafion content, were chosen and added to the final Nafion solution. The mixture obtained was casted on a Petri dish and dried at 80 °C. In order to improve the thermal stability and robustness of the membranes, dry membranes were extracted and hot-pressed at 50 atm, 175 °C, for 15 min. The membranes were activated and purified in boiling 3 wt.% hydrogen peroxide (H_2O_2, 34.5–36.5%, Sigma-Aldrich, St. Louis, MO, USA), H_2SO_4 (0.5 M) and distilled water. Composite membranes were compared to plain Nafion systems prepared with the same procedure. All samples were stored in bi-distilled water. Membranes containing 5 and 10 wt.% of the inorganic filler are labeled in the text as M5 and M10, respectively, while the undoped membrane is referred to as N.

The mechanical properties of the membranes were measured by means of a DMA 8000 (Perkin Elmer Waltham, MA, USA) in the so-called "tension configuration", on small membrane pieces that were 4–6 mm wide, 10–12mm long, and 0.10–0.15mm thick [18–21]. The samples were cut from the various membranes dried in an oven, at 80 °C (dry samples), or immersed in bi-distilled water, at room temperature (wet samples). In this latter case, to prevent the release of water, the samples were very quickly mounted into the DMA apparatus and measured. The storage modulus, M, and the elastic energy dissipation, tanδ, were measured at 1 and 10 Hz, as a function of temperature between 20 and 190 °C, with a scan rate of 4 °C/min.

DSC experiments were carried out using a DSC821 instrument (Mettler-Toledo, Zaventem, Belgium), under nitrogen (N_2) flux (60 mL/min), in a temperature range between 30 and 150 °C, at a scan rate of 20 °C/min. Before DSC measurements, membrane samples were equilibrated at 100% relative humidity (RH) for two weeks. By using the STARe software, the determination of both the T_{onset} and the enthalpy values associated with the thermal transition were evaluated. In particular, the T_{onset} was defined as the intersection of the tangent of the peak with the extrapolated baseline, whereas the peak area was proportional to the enthalpy of the thermal event.

Thermal gravimetric (TG) analysis was performed on dry samples, with a TGA/SDTA851 (Mettler-Toledo, Zaventem, Belgium), under air (80 mL/min), in a temperature range between 25 and 550 °C. Prior to measurements, the samples were dried at 80 °C, under vacuum, overnight.

Dielectric measurements were performed using a NovocontrolGmBH broadband dielectric spectrometer (Montabaur, Germany), equipped with a QuatroCryosystem temperature-control unit. The membranes were placed between two carbon electrodes and then between two gold-plated electrodes (diameter of 10 mm), under humidity condition (i.e., we kept a water reservoir under the bottom electrode. The spectra were measured in the frequency range from 10^{-1} to 10^7 Hzand at different temperatures, with the following temperature sequence: 20 °C → 80 °C → 110 °C → 80 °C → 20 °C.

3. Results and Discussion

3.1. Differential Scanning Calorimetry and Thermal Gravimetric Analysis

The DSC curves of all membranes are displayed in Figure 1. In the investigated temperature range, a main peak is observed around 100 °C, assigned to an order–disorder transition of the ionic clusters in Nafion [22]. The enthalpy value, calculated by integrating the DSC peak, associated to this endothermic phenomenon, was evaluated and is reported in Table 1. In accordance to the literature [22], only water associated with Nafion hydrophilic groups contributed to this thermal transition. In particular, the enthalpy value increased with an increasing degree of hydration of the polymer, leading to a major organization of ionic clusters and more cohesive interactions. At the same time, the change in temperature of the transition peak is attributed to a plasticizing effect of water, for which a shift toward lower temperatures may correspond to higher hydration levels. However, in our case, the change in the enthalpy values is much more significant than that in T_{onset}.

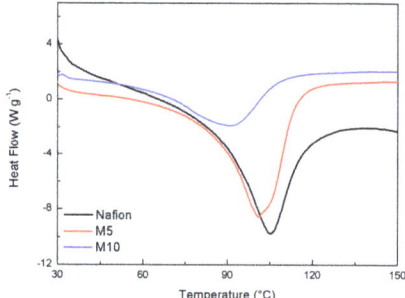

Figure 1. DSC response of the hydrated membrane samples.

Table 1. ΔH and T_{onset} associated to the thermal transition observed by DSC.

Samples	T_{onset} (°C)	ΔH (J·g^{-1})
N	82	738
M5	81	747
M10	55	380

Among all samples, the composite membrane M5 (containing 5 wt.% of the CaTiO$_{3-\delta}$ additive) shows a higher ΔH value than both M10 (10 wt.% of CaTiO$_{3-\delta}$) and, to a lower extent, N (plain Nafion) samples. The addition of CaTiO$_{3-\delta}$ particles caused an increase in the water content, even though the M10 sample, having the highest concentration of additive, displayed the lowest water affinity, possibly due to phase segregation and a non-optimized distribution of the inorganic additive [23,24], which can prevent the motions of the segments among the fluorocarbon backbone to restrict the water release.

The TGA curves obtained for the three membranes are shown in Figure 2 (panel a). The thermal decomposition of the Nafion membrane occurs in three main steps. The first decomposition is associated with desulfonation of the side-chain of the polymer; the second and the third transitions, occurring in the range 350–450 °C, are related to side-chains and perfluorinated backbone decompositions [25].

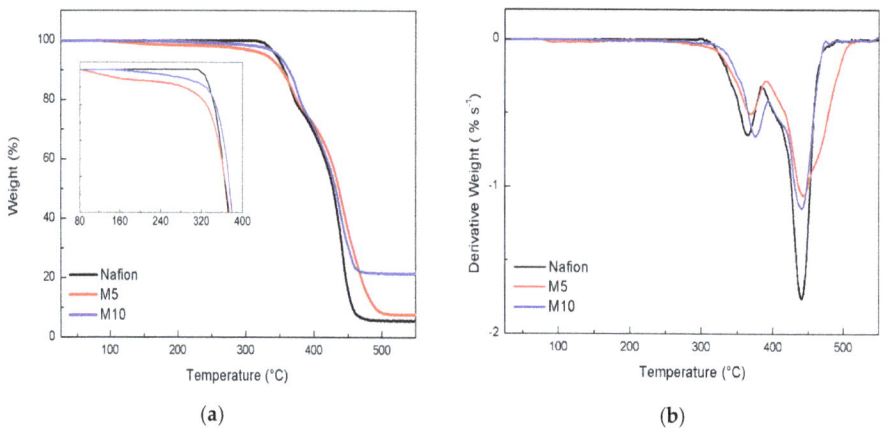

Figure 2. (a) TG and DTG curves (b) recorded for the dry membranes.

The TGA profiles show that all the membranes are thermally stable up to 300 °C, even though, compared to plain Nafion, the two composite samples exhibit slightly higher decomposition temperatures, as better shown by the derivative curves (see DTG curves in Figure 2b). In particular, the last thermal process looks broader and shifted to higher temperatures for the M5 sample, suggesting a stabilizing interaction between the filler and the Nafion matrix. Moreover, as shown in the inset of TGA profiles (see Figure 2a), the composite membranes exhibit a smooth mass loss in a lower temperature range (below 300 °C), most likely due to traces of surface water still present in the membrane after the drying treatment carried out at 80 °C prior to the measurement. This can be explained in terms of extra hygroscopicity induced by the inorganic fillers. The presence of surface water looks to be more evident in M5. This could be explained by considering an optimal concentration/dispersion of the perovskite additive, able to hold water during the drying step and to release it gradually at higher temperature. The greater slope observed in M5, as compared to M10 (see inset of Figure 2a), suggests a higher content of surface water. Considering the final weights after TGA analysis, it is clear that the M10 membrane leaves a higher amount of residual weight, in respect to the other samples. This evidence could be interpreted by assuming a strong interaction between the Nafion polymer and the perovskite particles, for which the removal of the decomposed Nafion products at a high temperature is more difficult.

3.2. Dynamical Mechanical Analysis

Figure 3 reports the storage modulus and the elastic energy dissipation of the dry membranes measured at two fixed frequencies (1 and 10 Hz), during heating between room temperature and 180 °C, with arate of 4 °C/min. In agreement with DSC measurements, the Nafion membrane shows a relaxation around 110 °C, indicated by the occurrence of an intense peak in tanδ and a two-order magnitude drop in the modulus. This relaxation, usually indicated as α-relaxation, was already largely reported [18,19] and corresponds to the glass transition of the hydrophilic domains (polar regions) of Nafion [26].

In the composite membraneM5, the α-relaxation is slightly shifted to higher temperatures compared to the pure Nafion sample, and this shift is even more clear when increasing the amount of filler, since, for the M10 samples, the relaxation is observed at around 130 °C. At room temperature, the modulus values of the three samples are close, while at higher temperature, the composite membranes present slightly higher modulus values, thus confirming the reinforcing action of the filler. In particular, above 90 °C the modulus of the M10 membrane is the highest. Indeed, at room temperature, the modulus values could be affected by some undesired water contamination occurring during sample loading, while on heating above 100 °C, these effects should be suppressed. An increase of the modulus in a Nafion membrane, with an addition of a filler like SnO_2 nanoparticles [27] and sulphated SnO_2 ceramic nano-powders [19], was already observed.

The modulus and the elastic-energy dissipation of the wet membranes measured at a fixed frequency of 1 Hz during heating between room temperature and180 °C, with a rate of 4 °C/min, are reported in Figure 4. The α-relaxation is shifted to slightly higher temperatures compared to the dry membranes, and the highest temperature for the maximum of the energy dissipation is observed for the wet M10 membranes, confirming that the faded filler increases the relaxation temperature. On cooling (Figure 4), the peak associated to the α-relaxation presents its maximum at about 90 °C for the pure Nafion and the M5 membranes. In both membranes the α-relaxation displays a clear thermal hysteresis between heating and subsequent cooling, as already reported for similar systems [18,19].

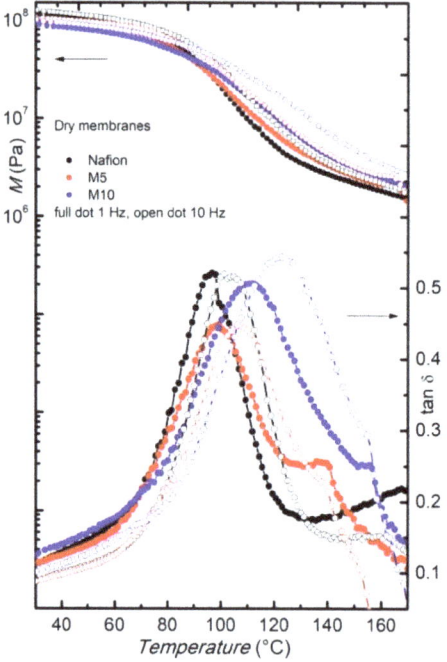

Figure 3. Modulus and elastic energy dissipation of pure and dry Nafion (black) and composite membranes M5 (red) and M10 (blue), measured on heating, at two frequencies, i.e., $f = 1$ Hz (full dots) and $f = 10$ Hz (open dots).

Contrarily, the peak displayed by the M10 membrane does not present a significant temperature shift when measured on cooling. For comparison, the curves (both storage modulus and tanδ) measured on cooling for the dry samples are also reported in Figure 4. The curves measured in the cooling run for the Nafion and the M5 membranes, starting from their wet state, are close to the curves measured on cooling for the corresponding dry samples. Conversely, the tanδ measured on cooling for the M10 membranes, starting from the wet state, displays the α-relaxation at the same temperature at which it is detected in the heating run, and well above the temperature at which it appears in the spectrum measured on cooling, starting from the dry state. Indeed, it seems that, for this latter sample, the α-relaxation shows a very small thermal hysteresis between heating and cooling, regardless of the water content, maybe due to some interactions between the higher amount of filler and the Nafion matrix. Moreover, with a higher amount of filler, the presence of water (i.e., at wet conditions) seems to increase, more remarkably, the temperature at which this relaxation occurs.

The modulus values of the wet Nafion and M5 samples measured at room temperature before heating are close to that displayed by the corresponding membranes in the dry state, while the modulus displayed by the wet M10 membrane is slightly lower than the one measured on the dry M10 membrane, suggesting that, in this latter case, water acts more remarkably as a plasticizer that decreases the stiffness of the membranes. However, when comparing the modulus value in the cooling run, where most of the water may have evaporated, the curves are close for the same kind of samples, independent of the initial state (wet or dry). This confirms that the thermal treatment completely eliminates the hydration history of the samples. A small difference can be noticed only for the M5 membranes, where the modulus measured in the cooling run, starting from the wet state, is slightly higher than those obtained when cooling the dry sample.

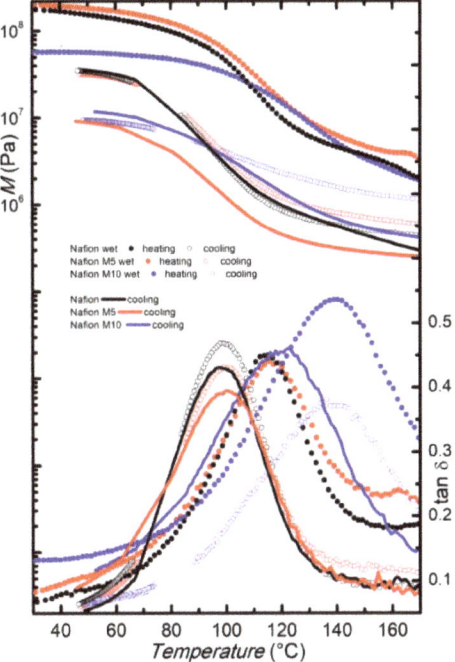

Figure 4. Modulus and elastic energy dissipation of pure Nafion (black) and composite membranes M5 (red) and M10 (blue) in the wet state, measured at $f = 1$ Hz, on heating (full dots) and subsequent cooling (open dots). Dry membranes measured at $f = 1$ Hz, on cooling (lines), are also reported for comparison.

3.3. Dielectric Spectroscopy Studies

Figure 5 shows the frequency dependence of the imaginary part of the permittivity (ε'') as a function of frequency for plain Nafion, M5, and M10, at a given temperature of 20 °C, during cooling.

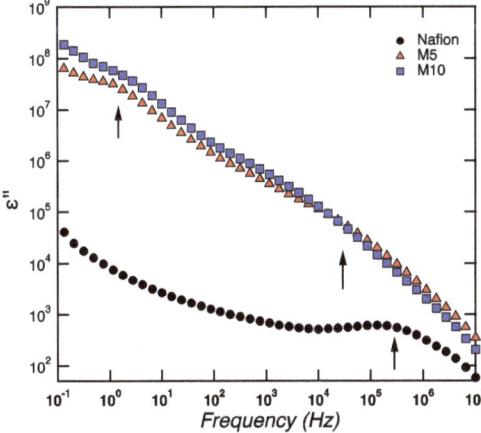

Figure 5. Frequency dependence of the imaginary part of the permittivity for Nafion, M5, and M10, at a given temperature of 20 °C, during cooling.

For samplesM5 and M10, two clear dielectric relaxation processes are observed at high and low frequencies. These relaxation processes, called β-relaxations (i.e.,$β_1$, $β_2$) and generally observed at a low temperature (i.e., lower than T_g), were attributed to conformational changes of the ether group bound to the backbone end of the side-chain or the ether group bound to the sulfonate end of the side-chain [28]. These β-relaxations are completely different from the main structural α-relaxation, which occurs at higher temperatures, around 110 °C, as revealed by DMA measurements, and is associated to the long-range movement of the fluorocarbon domains (i.e., to T_g). In the case of Nafion, one major β-relaxation process is observed at high frequencies, while, at the limit of the lowest frequency investigated, the shape of the curve is reminiscent of the tail of a β-relaxation (peaked at low frequencies outside the frequency window investigated).A similar behavior was found in Nafion membranes investigated by Di Noto et al. [28,29]. Both β-relaxations (i.e.,$β_1$, $β_2$) become faster with increasing temperature and move to higher frequencies, which implies that, at 80 °C, both are outside the experimental frequency window investigated (see Figure 6). This figure shows that the main structural α-relaxation is not observed within the studied frequency window, but two conductivities are detected instead. These are recognized by the typical slope of −1.

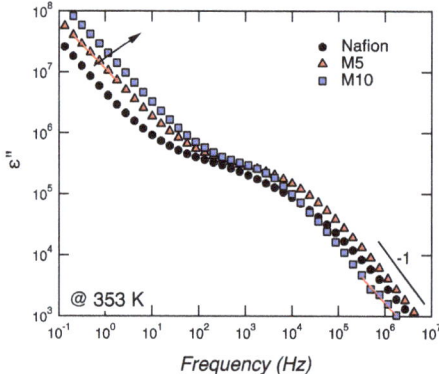

Figure 6. Frequency dependence of the imaginary part of the permittivity for Nafion, M5, and M10, at a given temperature of 80 °C, during heating.

We propose to assign the conductivity observed at low frequencies (i.e., in the range 10^{-1}–10^1 Hz) to localized and slow phenomena, whereas the conductivity at higher frequencies (i.e., in the range 10^4–10^6 Hz) is attributed to long-range bulk or dc conductivity. Both these two conductivities are associated to $ε''$, with a slope of −1, and become faster with an increasing concentration of $CaTiO_3$ (see arrow in Figure 6).

The conductivity curves recorded for all three membranes (Nafion, M5, and M10) and over the whole frequency window, as well as for all temperatures (i.e., 20, 80, and 110 °C) are shown in Figure 7, both for the heating and cooling scans.

Two characteristic ranges can be observed, at low (10^{-1}–10 Hz) and high (10^4–10^6 Hz) frequencies, a behavior similar to that described by Di Noto et al., in the investigation of dry and wet Nafion [28].

The conductivity at low frequencies (10^{-1}–10^1 Hz) shows a similar general trend for the three samples, increasing with increasing temperatures during heating and decreasing, with decreasing temperature during cooling. However, the conductivity at high frequencies (10^4–10^6 Hz) shows different behaviors in the three samples. For Nafion, the conductivity values are lowered by one order of magnitude after heating to 110 °C, which most likely reflects the dehydration of the membrane. For M5, the conductivity values during heating and cooling show a small but detectable change, with the conductivity increasing upon increased temperature. M5 also shows the highest conductivity, reaching a value at 110 °C of 3.1×10^{-3} S·cm^{-1} at 110 °C. For M10, there is no significant change in conductivity during heating and cooling, and a value about 1.2×10^{-3} S·cm^{-1} is measured. These trends are better visualized in Figure 8.

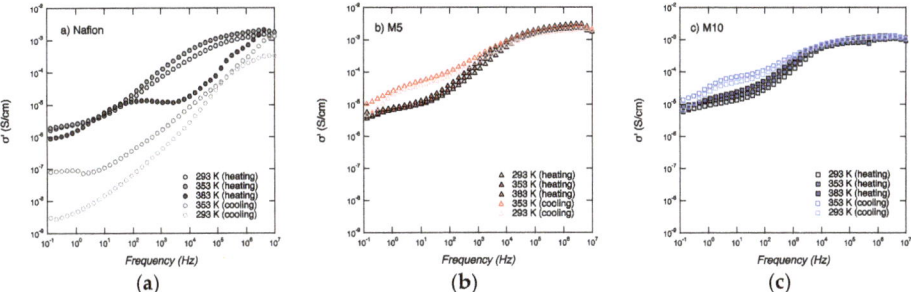

Figure 7. Conductivity (σ') spectra as a function of frequency for Nafion (**a**), M5 (**b**), and M10 (**c**), during heating and cooling.

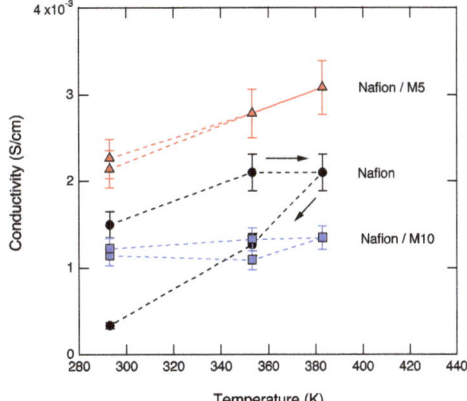

Figure 8. The dc conductivity as a function of temperature for Nafion, M5, and M10, during heating and cooling.

Table 2 shows the dependence of the dc conductivity on composition, for the representative temperature of 110 °C. The three samples all exhibit a reasonably high conductivity (~10^{-3} S·cm^{-1}), although M5 displays the highest value, indicating that a too-high concentration of CaTiO$_3$ can be detrimental for the conducting property. The conductivity displayed by M5 at 110 °C is higher than that of dry Nafion at the same temperature [28] and slightly lower than, but comparable to, that of Nafion containing sulfated zirconia as the filler (also added at 5 wt.%) [29].

Table 2. Conductivity values (σ) measured for different membranes at 110 °C.

Sample	wt.% CaTiO$_3$	σ (mS·cm^{-1})
N	0	2.1
M5	5	3.1
M10	10	1.3
Nafion (wet) [28]	-	10–100
Nafion (dry) [28]	-	0.0001–0.001
S-ZrO$_2$/Nafion (dry) [29]	-	1–10

The reported results indicate that composite membranes obtained by adding calcium titanate as filler in a Nafion matrix display interesting properties that ought to be considered for electrochemical applications, such as in PEMFCs. An improved water affinity and enhanced proton conductivity is observed for a low concentration of the filler (around 5 wt.%), whereas higher filler contents (about 10 wt.%) are not beneficial for the thermal and conducting performance of the membrane. This behavior was already reported for other types of filler in composite membranes [24]. In particular, previous results published by Chen et al. [30] show that Nafion membranes containing 5 wt.% of sulphated tin oxide had higher proton conductivity than both the undoped membrane and the membrane with 10 wt.% of filler. Moreover, filler loading around 4–5 wt.% was the most effective, also for Nafion-based composite membranes containing different metal oxide fillers, likely due to the more homogeneous dispersion of the additive within the polymer matrix [8,27]. The issues of inhomogeneous dispersion and non-optimized filler-to-polymer interactions appear to be particularly critical in this work due to the micrometric size of the calcium titanate particles [14], exceeding the dimension of the hydrophilic domains and, possibly, occluding them to some extent. Moreover, a too-high filler concentration could block the ionic channels and impede ionic motion, as also suggested for proton-conducting hybrid membranes containing protic ionic liquids and silica nanoparticles or mesoporous silica nanospheres [31].

4. Conclusions

In the present paper, a composite membrane based on a Nafion polymer matrix incorporating a $CaTiO_{3-\delta}$ additive is proposed and investigated. Different filler contents, namely two concentrations of 5 and 10 wt.% of the $CaTiO_{3-\delta}$ additive, with respect to the dry Nafion content, were considered. The membrane with the lower amount of additive displayed better properties in terms of both water affinity and conductivity. Indeed, our results suggest that a too-high content of additive can be detrimental for these particular properties. However, our results indicate that perovskite oxides can be useful as a water-retention and reinforcing additive in low-humidity proton-exchange membranes.

Author Contributions: Conceptualization, L.M., O.P., and M.A.N.; methodology, O.P., M.A.N., A.M., and A.P.; formal analysis, L.M., F.T., and K.E.; investigation, L.M., F.T., and K.E.; writing—original draft preparation, L.M., O.P., M.A.N., and A.M.; writing—review and editing, all the authors.

Funding: A.M. and K.E. acknowledge funding from the Swedish Foundation for Strategic Research (SSF, FFL-6 program, grant no FFL15-0092) and from the Knut & Alice Wallenberg Foundation (Academy Fellows program, grant no. 2016–0220).

Acknowledgments: L.M. acknowledges Erasmus+ Programme—Student Mobility for Traineeship—Project Unipharma-Graduates 2018/2019 of Sapienza University of Rome for the travel grant allowing the traineeship at Chalmers University of Technology, Gothenburg, Sweden.

Conflicts of Interest: The authors declare no conflicts of interest.

References

1. Rosli, R.E.; Sulong, A.B.; Daud, W.R.W.; Zulkifley, M.A.; Husaini, T.; Rosli, M.I.; Majlan, E.H.; Haque, M.A. A review of high-temperature proton exchange membrane fuel cell (HT-PEMFC) system. *Int. J. Hydrogen Energy* **2017**, *42*, 9293–9314. [CrossRef]
2. Bose, S.; Kuila, T.; Nguyen, T.X.H.; Kim, N.H.; Lau, K.T.; Lee, J.H. Polymer membranes for high temperature proton exchange membrane fuel cell: Recent advances and challenges. *Prog. Polym. Sci.* **2011**, *36*, 813–843. [CrossRef]
3. Kundu, S.; Simon, L.C.; Flowler, M.; Grot, S. Mechanical Properties of Nafion™ Electrolyte Membranes under Hydrated Conditions. *Polymer* **2005**, *46*, 11707–11715. [CrossRef]
4. Peighambardoust, S.J.; Rowshanzamir, S.; Amjadi, M. Review of the proton exchange membranes for fuel cell applications. *Int. J. Hydrogen. Energy* **2010**, *35*, 9349–9384. [CrossRef]
5. Brutti, S.; Scipioni, R.; Navarra, M.A.; Panero, S.; Allodi, V.; Giarola, M.; Mariotto, G. SnO_2-Nafion® nanocompositepolymerelectrolytes for fuelcellapplications. *Int. J. Nanotechnol.* **2014**, *11*, 882–896. [CrossRef]

6. Allodi, V.; Brutti, S.; Giarola, M.; Sgambetterra, M.; Navarra, M.A.; Panero, S.; Mariotto, G. Structural and spectroscopic characterization of a nanosized sulfated TiO_2 filler and of nanocomposite nafion membranes. *Polymers* **2016**, *8*, 68. [CrossRef] [PubMed]
7. Navarra, M.A.; Croce, F.; Scrosati, B. New, high temperature superacid zirconia-doped Nafion™ composite membranes. *J. Mater. Chem.* **2007**, *17*, 3210–3215. [CrossRef]
8. Siracusano, S.; Baglio, V.; Navarra, M.A.; Panero, S.; Antonucci, V.; Aricò, A.S. Investigation of Composite Nafion/Sulfated Zirconia Membrane for Solid Polymer Electrolyte Electrolyzer Applications. *Int. J. Electrochem. Sci.* **2012**, *7*, 1532–1542.
9. Goodenough, J.B.; Zhou, J.S. *Localized to Itinerant Electronic Transition in Perovskite Oxides*; Springer: Berlin, Germany, 2001; Volume 98, pp. 17–113.
10. Fabbri, E.; Mohamed, R.; Levecque, P.; Conrad, O.; Kötza, R.; Schmidt, T.J. Unraveling the Oxygen Reduction Reaction Mechanism and Activity of d-Band Perovskite Electrocatalysts for Low Temperature Alkaline Fuel Cells. *ECS Trans.* **2014**, *64*, 1081–1093. [CrossRef]
11. Du, J.; Zhang, T.; Cheng, F.; Chu, W.; Wu, Z.; Chen, J. Nonstoichiometric Perovskite $CaMnO_{3-\delta}$ for Oxygen Electrocatalysis with High Activity. *Inorg. Chem.* **2014**, *53*, 9106–9114. [CrossRef]
12. Li, Q.; Yin, Q.; Zheng, Y.S.; Sui, Z.J.; Zhou, X.G.; Chen, D.; Zhu, Y.A. Insights into Hydrogen Transport Behavior on Perovskite Surfaces: Transition from the Grotthuss Mechanism to the Vehicle Mechanism. *Langmuir* **2019**, *35*, 9962–9969. [CrossRef] [PubMed]
13. Stølen, S.; Bakkenw, E.; Mohn, C.E. Oxygen-deficient perovskites: Linking structure, energetics and ion transport. *Phys. Chem. Chem. Phys.* **2006**, *8*, 429–447. [CrossRef] [PubMed]
14. Mazzapioda, L.; Lo Vecchio, C.; Paolone, A.; Aricò, A.S.; Baglio, V.; Navarra, M.A. Enhancing Oxygen Reduction Reaction Catalytic Activity Using $CaTiO_{3-\delta}$ Additive. *Chem. Electro. Chem.* **2019**. [CrossRef]
15. Siracusano, S.; Baglio, V.; Nicotera, I.; Mazzapioda, L.; Aricò, A.S.; Panero, S.; Navarra, M.A. Sulfated titania as additive in Nafion membranes for water electrolysis applications. *Int. J. Hydrogen Energy* **2017**, *42*, 27851–27858. [CrossRef]
16. Branchi, M.; Sgambetterra, M.; Pettiti, I.; Panero, S.; Navarra, M.A. Functionalized Al_2O_3 particles as additives in proton-conducting polymer electrolyte membranes for fuel cell applications. *Int. J. Hydrogen Energy* **2015**, *40*, 14757–14767. [CrossRef]
17. Nicotera, I.; Kosma, V.; Simari, C.; Ranieri, G.A.; Sgambetterra, M.; Panero, S.; Navarra, M.A. An NMR study on the molecular dynamic and exchange effects in composite Nafion/sulfated titania membranes for PEMFCs. *Int. J. Hydrogen Energy* **2015**, *40*, 14651–14660. [CrossRef]
18. Teocoli, F.; Paolone, A.; Palumbo, O.; Navarra, M.A.; Casciola, M.; Donnadio., A. Effects of water freezing on the mechanical properties of Nafion membranes. *J. Polym. Sci. Pol. Phys.* **2012**, *50*, 1421–1425. [CrossRef]
19. Scipioni, R.; Gazzoli, D.; Teocoli, F.; Palumbo, O.; Paolone, A.; Ibris, N.; Brutti, S.; Navarra, M.A. Preparation and characterization of nanocomposite polymer membranes containing functionalized SnO_2 additives. *Membranes* **2014**, *4*, 123–142. [CrossRef]
20. Trequattrini, F.; Palumbo, O.; Vitucci, F.M.; Miriametro, A.; Croce, F.; Paolone, A. Hot pressing of electrospun PVdF-CTFE membranes as separators for lithium batteries: A delicate balance between mechanical properties and retention. *Mater. Res.* **2018**, *21*. [CrossRef]
21. Paolone, A.; Palumbo, O.; Teocoli, F.; Cantelli, R.; Hassoun, J. Phase transitions in polymers for lithium batteries. *Solid State Phenom.* **2012**, *184*, 351–354. [CrossRef]
22. Lage, L.G.; Delgado, P.G.; Kawano, Y. Thermal stability and decomposition of Nafion® membranes with different cations. *J. Therm. Anal. Calorim.* **2004**, *75*, 521–530. [CrossRef]
23. Sgambetterra, M.; Brutti, S.; Allodi, V.; Mariotto, G.; Panero, S.; Navarra, M.A. Critical Filler Concentration in Sulfated Titania-Added Nafion™ Membranes for Fuel Cell Applications. *Energies* **2016**, *9*, 272. [CrossRef]
24. Mazzapioda, L.; Panero, S.; Navarra, M.A. Polymer Electrolyte Membranes Based on Nafion and a Superacidic Inorganic Additive for Fuel Cell Applications. *Polymers* **2019**, *11*, 914. [CrossRef] [PubMed]
25. Di Noto, V.; Boaretto, N.; Negro, E.; Giffin, G.A.; Lavina, S.; Polizzi, S. Inorganic-organic membranes based on Nafion, $[(Zr_{O2})*(Hf_{O2})_{0.25}]$ and $[(Si_{O2})*(Hf_{O2})_{0.28}]$. Part I: Synthesis, thermal stability and performance in a single PEMFC. *Int. J. Hydrogen. Energy* **2012**, *37*, 6199–6214. [CrossRef]
26. Kyu, T.; Eisenberg, A. *Perfluorinated Ionomer Membranes*; Eisenberg, A., Yeager, H.L., Eds.; ACS Symposium Series 180; American Chemical Society: Washington, DC, USA, 1982; pp. 79–84.

27. Nørgaard, C.F.; Nielsen, U.G.; Skou, E.M. Preparation of Nafion 117™-SnO$_2$ Composite Membranes using an Ion-Exchange Method. *Solid State Ionics* **2012**, *213*, 76–82. [CrossRef]
28. Di Noto, V.; Piga, M.; Pace, G.; Negro, E.; Lavina, S. Dielectric Relaxations and Conductivity Mechanism of Nafion: Studies Based on Broadband Dielectric Spectroscopy. *ECS Trans.* **2008**, *16*, 1183–1193.
29. Guinevere, A.G.; Piga, M.; Lavina, S.; Navarra, M.A.; D'Epifanio, A.; Scrosati, B.; Di Noto, V. Characterization of sulfated-zirconia/Nafion® composite membranes for proton exchange membrane fuel cells. *J. Power. Source* **2012**, *198*, 66–75.
30. Chen, F.; Mecheri, B.; D'Epifanio, A.; Traversa, E.; Licoccia, S. Development of Nafion/Tin oxide composite MEA for DMFC applications. *Fuel Cells* **2010**, *10*, 790–797. [CrossRef]
31. Lin, B.; Cheng, S.; Qiu, L.; Yan, F.; Shang, S.; Lu, J. Protic Ionic Liquid-Based Hybrid Proton-Conducting Membranes for Anhydrous Proton Exchange Membrane Application. *Chem. Mater.* **2010**, *225*, 1807–1813. [CrossRef]

© 2019 by the authors. Licensee MDPI, Basel, Switzerland. This article is an open access article distributed under the terms and conditions of the Creative Commons Attribution (CC BY) license (http://creativecommons.org/licenses/by/4.0/).

Article

LFP-Based Gravure Printed Cathodes for Lithium-Ion Printed Batteries

Maria Montanino [1,*], Giuliano Sico [1], Anna De Girolamo Del Mauro [1] and Margherita Moreno [2]

[1] ENEA Italian National Agency for New Technologies, Energy and Sustainable Economic Development, SSPT-PROMAS-NANO, 80055 Portici, Italy; giuliano.sico@enea.it (G.S.); anna.degirolamo@enea.it (A.D.G.D.M.)
[2] ENEA Italian National Agency for New Technologies, Energy and Sustainable Economic Development, DTE-PCU-SPCT, 00123 Roma, Italy; margherita.moreno@enea.it
* Correspondence: maria.montanino@enea.it

Received: 30 April 2019; Accepted: 5 June 2019; Published: 7 June 2019

Abstract: Printed batteries have undergone increased investigation in recent years because of the growing daily use of small electronic devices. With this in mind, industrial gravure printing has emerged as a suitable production technology due to its high speed and quality, and its capability to produce any shape of image. The technique is one of the most appealing for the production of functional layers for many different purposes, but it has not been highly investigated. In this study, we propose a LiFePO$_4$ (LFP)-based gravure printed cathode for lithium-ion rechargeable printed batteries and investigate the possibility of employing this printing technique in battery manufacture.

Keywords: gravure printing; printed batteries; printed cathode; lithium batteries; multilayer

1. Introduction

The use of printing techniques as a low cost production method for creating layers of different functional materials has recently undergone increased investigation in many fields. Compared with coating techniques, printing allows greater control of the characteristics of the layer, the possibility to realize any desired shape and pattern, and it also delivers a higher production speed. Among these printing techniques, the gravure technique is the most commonly used for the production of magazines and flexible packaging because of its ability to couple high throughput (speed up to 400 m/min^{-1}) and high quality (resolution 0.1 µm). Gravure is considered to be the most promising technique for producing thin layers (0.05–10 µm) of different functional materials [1]. The use of such conventional roll-to-roll industrial printing techniques allows the manufacture of low cost flexible structures and devices at high volume [2,3] in a one-step direct deposition process, which is suitable for patterning realization and large area production under ambient conditions, coupled with a minimal waste of energy, time, and materials [4–6]. With this aim, in the last few years organic materials, such as polymers and conductive polymers, have been successfully gravure printed in our laboratories to be employed in the field of optoelectronic [7–10]. More recently, this technique has also been demonstrated to be suitable for inorganic materials such as ceramics, offering the possibility to tailor the layer's characteristics through the modulation of the printing parameters [11]. The level of control of particle deposition is high enough to allow an innovative method of oxide sintering at low temperatures under pressure-less conditions [12].

In this paper, we demonstrate that it is possible to employ the gravure printing technique in the field of printed batteries. Printed batteries are thin batteries used in portable electronic devices, and their use is becoming more and more widespread in our daily lives [13]. All such devices (e.g., wearable, beauty, and biomedical) need only a small specific capacity (5–10 mAh·cm^{-2}), which has to be provided by a thin and customizable battery with a volume below 10 mm^3 in order for it to be perfectly integrated into the device. To date, industrially produced printed batteries are mostly not rechargeable [13].

Despite its possible advantages, the use of gravure printing for the production of printed batteries has not been well reported in scientific literature [14]. This is mainly due to the requirement of having to use low viscosity inks in order to achieve adequate thickness, particularly in electrodes, for achieving adequate capacities [15]. In addition, the possible contamination of materials from the printing cylinder (steel, copper, or chromium) limits the ink formulation. Finally, while the possibility to print polymers and inorganic materials separately has been demonstrated, the possibility to print such materials together in a homogenous composite structure remains a challenge.

In this study, gravure printing is used to produce $LiFePO_4$ (LFP)-based cathodes for rechargeable lithium-ion batteries using a multilayer approach. The well-known LFP was chosen as a reference. Moreover, in accordance with the most recent research regarding green aspects of component preparation, the cathodes were prepared using a water soluble sodium carboxy methyl cellulose (CMC) binder, which was deposited by water based ink solution.

2. Materials and Methods

Suitable inks were prepared for the gravure printing of the cathodes, with a fixed percentage of the solid component and a variable solvent content. The materials involved and their proportions were as follows: $LiFePO_4$ (LFP) (Sigma–Aldrich, Milan, Italy) as the active material (84%), super P (Thermofisher, Karlsruhe, Germany) as the conductive carbon (10%), and sodium carboxy methyl cellulose (CMC) (Panreac Quimica sa., Barcelona, Spain) as the binder (6%). The solvent used was a mixture of water and 2-propanol (80–20 wt%). The cathodes were constructed through a multilayer deposition of 3 layers (3L) or 5 layers (5L). The first layer was deposited by ink containing 23% by weight of dry content. The second layer was deposited by adding 10% of the mixed solvent to the ink used for the first layer. The successive layers were printed, adding a further 5% of the mixed solvent in each step. The solid content of each layer is reported in Table 1. The cathodic layers were deposited on aluminum foils (Sigma–Aldrich) using a commercial lab-scale IGT G1-5 gravure printer (IGT, Alemere, Netherlands) equipped with a cylinder with a line density of 40 lines/cm, stylus angle of 120°, cell depth of 72 µm, and screen angle of 53°. Each layer, and the finished cathodes, were dried at 130 °C. No final calendering was performed on the printed cathodes. After preliminary tests, the best printing conditions were found to be a printing force of 500 N at a speed of 36 m/min. The printing conditions were kept constant for all the printed layers. The electrical conductivity of the printed layers was verified by sheet resistance measurements performed by a four points probe instrument (Resistest RT 8A coupled with Resistage RG 8 supplied by Napson, Korea). The thickness and surface roughness of the printed samples were investigated by interferometry-based optical profilometer (Talysurf CCI HD, Taylor Hobson, Leicester, UK). The reported values represent the average obtained by several measurements and have a standard deviation of about 10%. The root mean square surface roughness was obtained according to the ISO 25178 standard. The morphology of the printed cathodes was also investigated through scanning electron microscopy (1530, LEO Elektronenmikroskopie GmbH, Oberkochen, Germany). The 3L and 5L cathodes were cut into discs of 14 mm diameter and tested in cells against lithium metal foil discs of 12 mm diameter. The separator was a glass fiber disc and the electrolyte used was an LP30 battery grade (Sigma–Aldrich) (1 M solution $LiPF_6$ in a 1:1 by volume mixture of ethylene carbonate and diethyl carbonate (EC:DEC, 1:1)). Galvanostatic cycling measurements were performed on the cells by a Maccor 4000 at 20 °C, at a fixed 0.1 C, and then at increasing C-rates.

Table 1. Layer by layer characteristics of the gravure printed cathodes.

Layer n.	Ink dry Content (%)	Overall Active Material (g cm^{-2})	Overall Thickness (μm)	Surface Roughness (nm)
1	23	9 10^{-5}	1.7	1
2	21	2 10^{-4}	3.2	1.5
3	20	4 10^{-4}	4.6	1.8
4	19	5 10^{-4}	5.9	2.3
5	18	5 10^{-4}	7.1	2.5

3. Results and Discussion

The gravure printing process consists of the fluid transfer of low-viscosity ink from the micro-engraved cells of a printing cylinder directly onto a flexible substrate through the pressure of a rubber cylinder as depicted in Figure 1. The desired geometry/patterning is obtained by engraving it onto the printing cylinder.

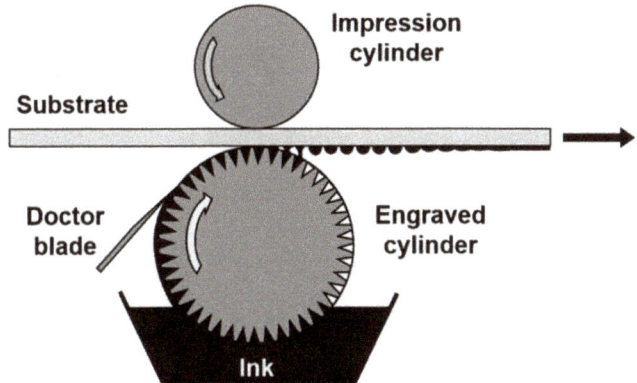

Figure 1. Schematic of the gravure printing process.

Several physical parameters relating to the materials, such as the ink and the substrate, are important in the gravure printing production quality; the ink viscosity, its rheological behavior, the surface tension/surface energy, the solvent evaporation rate, and the substrate porosity and smoothness. Moreover, process parameters such as cell geometry and density, printing pressure, and speed also play an important role on the final results. Although it may appear to be a relatively simple process, gravure printing has a complex multi-physical nature involving a series of sub-processes (inking, doctoring, transfer, spreading, and drying), each with its ideal operating regime, and each one determining the final quality of the printed product. In addition, an important issue is the formulation of low viscosity inks (1–100 mPa·s) [16] suitable for gravure printing that are able to realize proper functional layers. Thus, in order to prepare the inks, a large quantity of solvent is required. In this work, in order to develop a sustainable process, a water soluble CMC was used as a binder, and for this reason a mixture of water and 2-propanol was used in the ink formulation. The 2-propanol played the role of improving the ink printability, decreasing the surface tension due to the use of water, and improving the ink wettability of both the substrate and the printing cylinder. Taking into account all such matters, several preliminary tests were carried out to identify the best ink composition and process parameters, and the results are reported in the experimental section. Composite cathodic materials were successfully gravure printed onto aluminum foils and demonstrated good printability. To increase the mass loading, a multilayer approach was used. Up to five layers were overlapped using inks at decreasing dry content levels, keeping all the other printing parameters (cylinder, speed, pressure, drying temperature) constant, which benefited the overall production process. The multilayer was created by stacking at increasing solvent amounts in order to improve the distribution. This approach has been proved as the best way to lay down the consecutive printed layers [8]. The layer by layer characteristics of the printed cathodes are reported in Table 1. The mass loading of the active LFP material increased until the third layer. When adding another two layers (up to five), the increase in the mass loading and the thickness of the printed layers was poor due to the decreasing dry content of the inks. It had been expected that the increase in the solvent content of the inks on each of the layers would restrain the surface roughness increase, but this effect was not observed. This is likely due to the starting size distribution of the active material, which was measured in microns. In addition, the SEM images in Figure 2 show worse distribution and a slightly higher inhomogeneity in the 5L cathode when compared with the 3L.

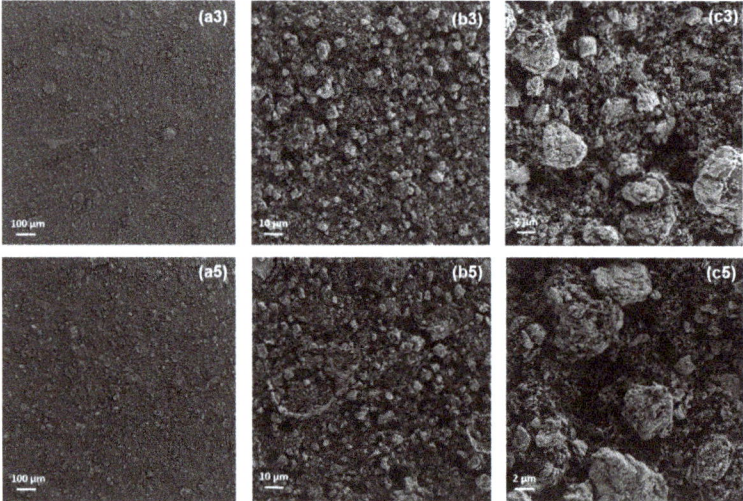

Figure 2. SEM images of the 3 layer (3L) (**a3,b3,c3**) and 5 layer (5L) (**a5,b5,c5**) gravure printed cathodes at different magnifications.

When the magnification of the cathode images is increased, it can be seen that the lower homogeneity of the 5L appears to be caused by polymer segregation occurring between the CMC polymer and the LFP active material. This is probably due to the low affinity of the CMC versus the increasing 2-propanol content in the ink generating the formation of polymer domains into the printed layer, thus worsening the distribution of the components in the 5L cathode itself. Both the cathodes were tested in batteries against lithium metal. In Figure 3, examples of charge-discharge cycles obtained for the 3L and 5L cathodes are reported.

The galvanostatic profiles appear featureless and present a typical LFP plateau, which is flat around 3.4 V in both charge and discharge, showing a stable cyclability and specific capacities close to the theoretical one (170 mAh/g). This is especially true for the 3L cathode. These results demonstrate that the structure of the printed layer is suitable to be used as a cathode. The discharge specific capacities of the 3L and 5L cathodes are shown in Figure 4.

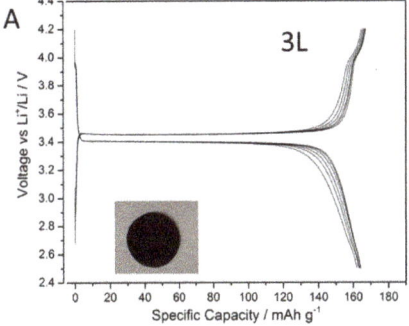
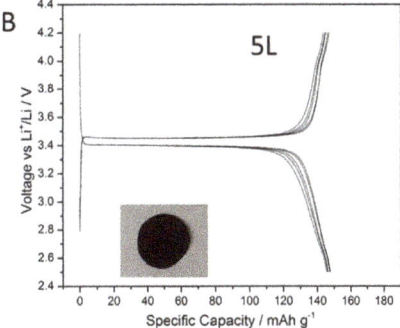

Figure 3. The galvanostatic profiles for cycles 5–10 obtained for the 3L (**A**) and the 5L (**B**) gravure printed cathode. The insets are photographs of the electrodes.

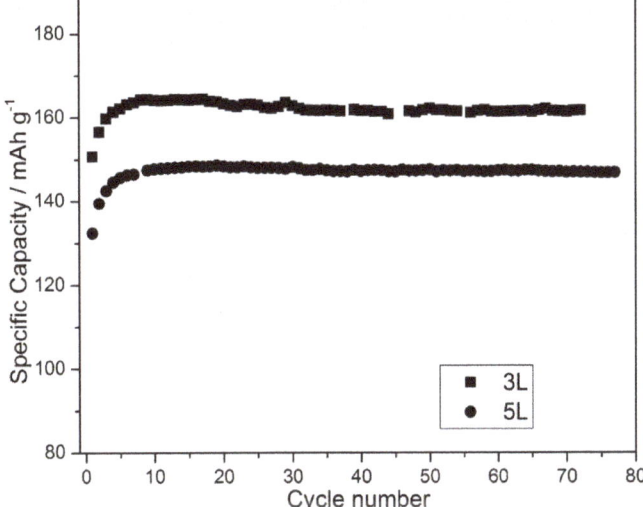

Figure 4. The discharge specific capacity vs. cycle number of the 3L and 5L gravure printed cathodes.

Sample 3L shows specific capacity values close to the theoretical ones, with a very high coulombic efficiency (>98%) for almost 100 cycles. After a few initialization cycles (<5) there is no capacity fading upon cycling at C/10. The same behavior can be observed in the 5L sample, but the values are 20 mAh lower. This provides important feedback regarding the production process. The higher homogeneity of the 3L cathode leads to higher efficiency in its behavior in batteries, thus demonstrating that the 3L cathode is better than the 5L cathode, especially when considering the necessary production steps. Since no calendering was performed on the printed cathodes, the positive battery tests suggest that using gravure printing would allow such post-process steps to be skipped, which would simplify the overall process. Cycling of the investigated cells was continued, and they showed good stability over time. Therefore, long life cyclability can be expected.

The charge and discharge capacities compared to the cycle numbers at increasing specific currents for the cells containing the 3L and 5L cathodes are reported in Figure 5. The Figure shows good stability of the cells at different rates and similar values for the charge and discharge capacities above 100 mAh·g^{-1}, even at 2 C-rate. However, the obtained mass loadings (see Table 1) are too low for practical applications, but they could be improved by increasing the thickness and density of the layers by decreasing the size and narrowing the size distribution of the active material. This would allow the printability of more concentrated inks, even when using different multilayer profiles. Moreover, a substrate pre-treatment, such as a commonly used corona discharge, would also improve the distribution of the solid on the substrate. Such changes would also improve the homogeneity of the printed electrode, further improving its performances. Furthermore, the use of a better performing active material than LFP would simplify the target achievement. Nevertheless, the good performance results, in terms of efficiency and reproducibility, of the printed cathodes in this study prove the feasibility of gravure printing in the field of printed batteries. This work may open the way for layer by layer device manufacturing using only gravure printing, which would potentially bring large advantages in terms of fast, easy, and low cost printed battery production.

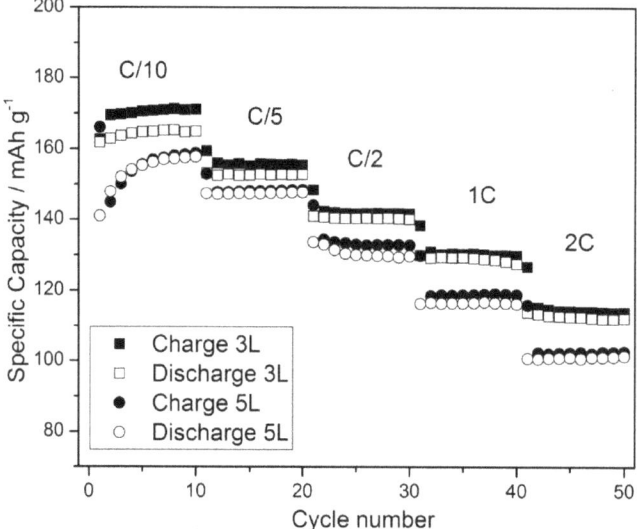

Figure 5. The charge-discharge specific capacity vs. the cycle number of the 3L and 5L gravure printed cathodes at increasing C-rates.

4. Conclusions

Thanks to a multilayer approach, the gravure printing technique led to the production of functional composite layers. The feasibility to gravure print cathodes for batteries has been demonstrated and, even with only a few overlapped layers, good performances was achieved. Keeping most of the printing parameters constant during the production process and skipping the calendering step allowed the manufacturing process to be simplified, which would make its industrial scaling easier. The performances of the printed cathodes could be improved by increasing the layer homogeneity by decreasing the size and the size distribution of the starting materials. These results open the way to the possibility of utilizing such techniques in future industrial production, especially in the field of printed batteries.

Author Contributions: Conceptualization, M.M. (Maria Montanino); Data curation, M.M. (Maria Montanino); Investigation, M.M. (Maria Montanino), G.S., A.D.G.D.M. and M.M. (Margherita Moreno); Methodology, M.M. (Maria Montanino) and G.S.; Writing—original draft, M.M. (Maria Montanino); Writing—review & editing, M.M. (Maria Montanino) and G.S.

Funding: Please add: Part of this work was funded by the research program "Ricerca di Sistema Elettrico" inside the "Accordo di Programma MiSE- ENEA".

Conflicts of Interest: The authors declare no conflict of interest.

References

1. Søndergaard, R.R.; Hosel, M.; Krebs, F.C. Roll-to-Roll Fabrication of Large Area Functional Organic Materials. *J. Polym. Sci. Part B Polym. Phys.* **2013**, *51*, 16–34. [CrossRef]
2. Tsay, C.-Y.; Wu, P.-W. Low temperature deposition of ZnO semiconductor thin films on a PEN substrate by a solution process. *Electron. Mater. Lett.* **2013**, *9*, 385–388. [CrossRef]
3. Kim, S.J.; Yoon, S.; Kim, H.J. Review of solution-processed oxide thin-film transistors. *Jpn. J. Appl. Phys.* **2014**, *53*, 02BA02. [CrossRef]
4. Puetz, J.; Aegerter, M.A. Direct gravure printing of indium tin oxide nanoparticle patterns on polymer foils. *Thin Solid Films* **2008**, *516*, 4495–4501. [CrossRef]
5. Alsaid, D.A.; Rebrosova, E.; Joyce, M.; Rebros, M.; Atashbar, M.; Bazuin, B. Gravure printing of ITO transparent electrodes for applications in flexible electronics. *J. Display Technol.* **2012**, *8*, 391–396. [CrossRef]
6. Khan, S.; Lorenzelli, L.; Dahiya, R. Technologies for printing sensors and electronics over large flexible substrates: A review. *IEEE Sens. J.* **2015**, *15*, 3164–3185. [CrossRef]
7. Montanino, M.; De Girolamo Del Mauro, A.; Tesoro, M.; Ricciardi, R.; Diana, R.; Morvillo, P.; Nobile, G.; Imparato, A.; Sico, G.; Minarini, C. Gravure-printed PEDOT:PSS on flexible PEN substrate as ITO-free anode for polymer solar cells. *Polym. Compos.* **2015**, *36*, 1104–1109. [CrossRef]
8. Sico, G.; Montanino, M.; De Girolamo Del Mauro, A.; Imparato, A.; Nobile, G.; Minarini, C. Effects of the ink concentration on multi-layer gravure-printed PEDOT:PSS. *Org. Electron.* **2016**, *28*, 257–262. [CrossRef]
9. Montanino, M.; Sico, G.; Prontera, C.T.; De Girolamo Del Mauro, A.; Aprano, S.; Maglione, M.G.; Minarini, C. Gravure printed PEDOT:PSS as anode for flexible ITO-free organic light emitting diodes. *Express Polym. Lett.* **2017**, *11*, 518–523. [CrossRef]
10. Sico, G.; Montanino, M.; De Girolamo Del Mauro, A.; Minarini, C. Improving the gravure printed PEDOT:PSS electrode by gravure printing DMSO post-treatment. *J. Mater. Sci. Mater. Electron.* **2018**, *29*, 11730–11737. [CrossRef]
11. Sico, G.; Montanino, M.; Prontera, C.T.; De Girolamo Del Mauro, A.; Minarini, C. Gravure printing for thin film ceramics manufacturing from nanoparticles. *Ceram. Int.* **2018**, *44*, 19526–19534. [CrossRef]
12. Sico, G.; Montanino, M.; Ventre, M.; Mollo, V.; Prontera, C.T.; Minarini, C.; Magnani, G. Pressureless sintering of ZnO thin film on plastic substrate via vapor annealing process at near-room temperature. *Scr. Mater.* **2019**, *164*, 48–51. [CrossRef]
13. Oliveira, J.; Costa, C.M.; Lanceros-Méndez, S. Printed Batteries: An Overview. In *Printed Batteries Materials, Technologies and Applications*, 1st ed.; Lanceros-Méndez, S., Costa, C.M., Eds.; John Wiley & Sons Ltd.: Chichester, UK, 2018; pp. 1–14.
14. Hwang, S.S.; Cho, C.G.; Park, K.-S. Stabilizing LiCoO$_2$ electrode with an overlayer of LiNi$_{0.5}$Mn$_{1.5}$O$_4$ by using a Gravure printing method. *Electrochem. Commun.* **2011**, *13*, 279–283. [CrossRef]
15. Rassek, P.; Wendler, M.; Krebs, M. Industrial Perspective on Printed Batteries. In *Printed Batteries Materials, Technologies and Applications*, 1st ed.; Lanceros-Méndez, S., Costa, C.M., Eds.; John Wiley & Sons Ltd.: Chichester, UK, 2018; pp. 185–192.
16. Krebs, F.C. Fabrication and processing of polymer solar cells: A review of printing and coating techniques. *Sol. Energy Mater. Sol. Cells* **2009**, *93*, 394–412. [CrossRef]

© 2019 by the authors. Licensee MDPI, Basel, Switzerland. This article is an open access article distributed under the terms and conditions of the Creative Commons Attribution (CC BY) license (http://creativecommons.org/licenses/by/4.0/).

MDPI
St. Alban-Anlage 66
4052 Basel
Switzerland
Tel. +41 61 683 77 34
Fax +41 61 302 89 18
www.mdpi.com

Membranes Editorial Office
E-mail: membranes@mdpi.com
www.mdpi.com/journal/membranes

www.ingramcontent.com/pod-product-compliance
Lightning Source LLC
LaVergne TN
LVHW070046120526
838202LV00101B/731